SCHAUM'S *Easy* OUTLINES

MOLECULAR AND CELL BIOLOGY

Other Books in Schaum's Easy Outlines Series Include:

SCHAUM'S *Easy* OUTLINES

MOLECULAR AND CELL BIOLOGY

BASED ON SCHAUM'S
Outline of Theory and Problems of
Molecular and Cell Biology
BY WILLIAM D. STANSFIELD, Ph.D.
JAIME S. COLOMÉ, Ph.D.
RAÚL J. CANO, Ph.D.

ABRIDGEMENT EDITOR
KATHERINE E. CULLEN, Ph.D.

SCHAUM'S OUTLINE SERIES
McGRAW-HILL

New York   Chicago   San Francisco   Lisbon   London   Madrid
Mexico City   Milan   New Delhi   San Juan
Seoul   Singapore   Sydney   Toronto

The *McGraw-Hill* Companies

WILLIAM D. STANFIELD was a faculty member of the Biological Sciences Department at California Polytechnic State University from 1963 until 1992 and is now Emeritus Professor. He holds a B.A. in agriculture, an M.A. in education, an M.S in genetics, and a Ph.D. in genetics from the University of California at Davis. He has written university-level textbooks in evolution and serology/immunology and is coauthor of a dictionary of genetics.

JAIME S. COLOMÉ teaches microbiology, genetics, cell physiology, and molecular biology at California Polytechnic State University. He holds a Ph.D. in molecular biology from the University of California at Santa Barbara. He is a coauthor of textbooks on microbiology as well as the author of several papers on the history of science.

RAÚL J. CANO is Professor of Microbiology in the Biological Sciences Department at the University of California, San Luis Obispo. Born in Cuba, he earned a B.S. and M.S. degrees from Eastern Washington University and his Ph.D. from the University of Montana. He has published more than 50 articles and has written two textbooks and a laboratory manual in microbiology.

KATHERINE E. CULLEN until recently taught biology at Transylvania University in Lexington, Kentucky, and is a teacher trainer for Kaplan Educational Services. She received a B.S. from the Michigan State University and a Ph.D. from Vanderbilt University. She has published several articles and was abridgement editor for *Schaum's Easy Outline: Biology* and *Schaum's Easy Outline: Biochemistry*.

1 2 3 4 5 6 7 8 9 0 DOC/DOC 0 9 8 7 6 5 4 3

ISBN 0-07-139881-3

Contents

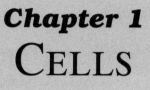

Chapter 1
CELLS

In This Chapter:

✔ *Introduction*
✔ *Cellular Organization*
✔ *Metabolism*
✔ *Reproduction*
✔ *Solved Problems*

Introduction

A **cell** is the smallest unit that exhibits all of the qualities associated with the living state. Cells must obtain energy from an external source to carry on such vital processes as growth, repair, and reproduction. All of the chemical and physical reactions that occur in a cell to support these functions constitute its **metabolism**. Metabolic reactions are catalyzed by **enzymes**. Enzymes are protein molecules  that accelerate biochemical reactions without being permanently altered or consumed in the process. The structure of each enzyme (or any other protein) is encoded by a segment of a deoxyribonucleic acid (DNA) molecule referred to as a **gene**.

Molecular and cell biology are the sciences that study all life processes within cells and at the molecular level. In doing so, these sci-

1

ences draw upon knowledge from several scientific disciplines, including biochemistry, cytology, genetics, microbiology, embryology, and evolution.

Cellular Organization

Structurally, there are two basic kinds of cells: **prokaryotic** and **eukaryotic**. Prokaryotic cells, including bacteria and archae, although far from simple, are generally much smaller and less complex structurally than eukaryotic cells. The major difference is that the genetic material (DNA) is not sequestered within a double-membrane structure called a **nucleus** (see Figure 1-1). In eukaryotes, a complete set of genetic instructions is found on the DNA molecules, which exist as multiple linear structures called **chromosomes** that are confined within the nucleus.

Eukaryotic cells also contain other membrane-bound **organelles** within their **cytoplasm** (the region between the nucleus and the plasma membrane). These subcellular structures vary tremendously in structure and function.

Most eukaryotic cells have **mitochondria,** which contain the enzymes and machinery for aerobic respiration and oxidative phosphorylation. Thus, their main function is generation of **adenosine triphosphate (ATP)**, the primary currency of energy exchanges within the cell. This organelle is bounded by a double membrane. The inner membrane, which houses the electron transport chain and the enzymes necessary for ATP synthesis, has numerous foldings called **cristae,** which protrude into the **matrix**, or central space. Mitochondria contain their own DNA and ribosomes, but most of their proteins are imported from the cytoplasm.

You Need to Know ✔

Mitochondria are nicknamed the "powerhouses" of the cell because of their role in ATP production.

Chloroplasts contain the photosynthetic systems for utilizing the radiant energy of sunlight and are found only in plants and algae. **Photo-**

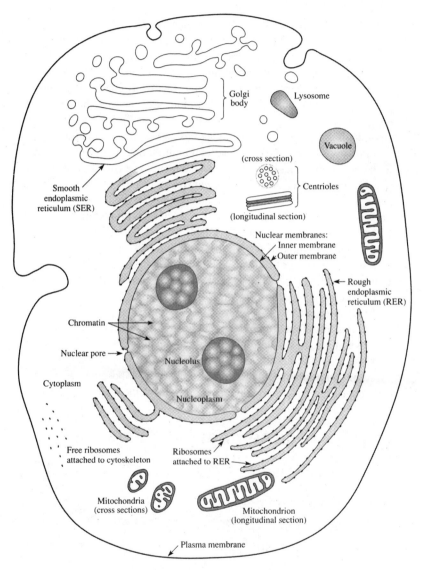

Figure 1-1 An animal cell.

synthesis is the process that converts light energy into the chemical bond energy of ATP, which in turn can be used to convert carbon dioxide (CO_2) and water (H_2O) into carbohydrates. Chloroplasts contain an internal system of membranes called **thylakoids**, a circular chromosome, and their own ribosomes. The flattened, vesicular thylakoids contain the chlorophyll pigments, the enzymes, and other molecules needed to harness light energy for conversion to chemical energy. Carbon fixation occurs in the **stroma**, the space between the thylakoids and the inner membrane.

Prokaryotic cells lack internal membranes, but photosynthetic bacteria contain invaginations of the plasma membrane called **mesosomes**.

Centrioles, located within the **centrosome**, are associated with the cell's polar regions, toward which the chromosomes migrate during cell division, and are found only in animal cells. The **endoplasmic reticulum** (ER) amplifies the surface area available for specialized biochemical reactions and the synthesis of certain types of proteins. The **Golgi complex** directs the transport of proteins and other biomolecules to specific locations within the cell. **Vacuoles** serve as storage compartments for food, water, or other molecules. Enzymes digest materials brought into the cell within **lysosomes**.

Ribosomes function in the manufacture of proteins. The ribosomes in prokaryotes are smaller than those found in the cytoplasm in eukaryotes, but are similar in size and structure to those found in the mitochondria and chloroplasts of eukaryotes. Eukaryotic ribosomes associated with the ER give it a granular appearance, hence the name **rough ER**.

Remember

Proteins that are:

1. membrane bound
2. secreted
3. compartmentalized

are synthesized on the ER.

Motility is accomplished by different means in prokaryotic and eukaryotic cells. Eukaryotic cells, such as amoebas and white blood cells, creep along substrates as an undulating mass of constantly changing morphology. This type of motion is achieved by a massive network of protein fibers, the **cytoskeleton**. Motile bacteria are usually propelled by one or more hairlike appendages called **flagella** that originate in the plasma membrane and rotate like propellar shafts (see Figure 1-2). These filaments are constructed of the protein **flagellin**. Some eukaryotic cells also have flagella, but they consist of bundles of microtubules made of **tubulin**, and they originate from a **basal body** in the cytoplasm. Eukaryotic flagella such as those in sperm tails bend back and forth in quasi-sinusoidal waves. Eukaryotic **cilia** are structurally similar but are much shorter, more numerous, and more rigid on the powerstroke. Some bacteria also have long hollow tubes called **pili** or **fimbriae** composed of a protein called **pilin**. These structures do not contribute to motility, but to the adhesiveness of bacteria and the facilitation of conjugation (see Chapter 7).

One of the distinguishing features between plants and animals is that plants and fungi have **cell walls** made of cellulose and chitin, respectively, but animal cells do not. Almost all bacteria have a rigid cell wall surrounding the plasma membrane, but it has a different structure than the plant cell wall and is composed of peptidoglycan. Some bacteria also have a polysaccharide **capsule** or a **glycocalyx** surrounding the cell wall.

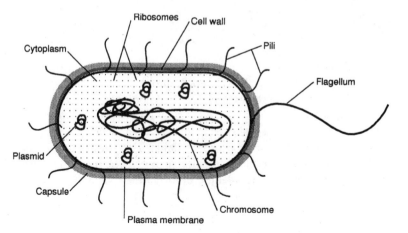

Figure 1-2 A bacterial cell.

These protect the bacteria from predatory cells and promote their attachment to various objects and to each other. Most eukaryotic cells also have a glycocalyx that covers the surface of the cell and promotes cell adhesions in the formation of specific tissues. In addition, many types of animal cells are surrounded by an **extracellular matrix,** which comprises a variety of proteins that give specific tissues their characteristic properties.

Metabolism

The two major carbon sources utilized by cells to synthesize organic molecules are (1) complex organic molecules, such as sugars and amino acids, and (2) single-carbon compounds, such as CO_2 or methane (CH_4). Cells that use CO_2 as their sole source of carbon are called **autotrophs,** and cells that require complex organic compounds are referred to as **heterotrophs.** Cells that can obtain energy from light are called **phototrophs,** and cells that require chemical energy are called **chemotrophs.**

Try it!

Distinguish between a photoautotroph and a photoheterotroph or a chemoautotroph and a chemoheterotroph.

Glycolysis is a nearly universal process in which the six carbon sugar glucose is anaerobically converted, through a series of enzymatically catalyzed steps in the **cytosol,** the fluid portion of the cytoplasm, into two molecules of the three carbon compound pyruvate. Two molecules of ATP are expended early on in glycolysis, but four more are generated later by substrate-level phosphorylation. Thus, there is a net production of two ATP molecules per molecule of glucose. In addition, two molecules of **nicotinamide adenine dinucleotide** (NAD) become reduced by gaining two electrons.

Remember Glycolysis!

Glucose + 2 NAD$^+$ + 2 ADP + 2 P$_i$ →
2 pyruvate + 2 ATP + 2 NADH + H$^+$

Either fermentation or respiration may follow glycolysis (see Figure 1-3). **Fermentation** is an oxygen-independent process, occuring in the cytosol, which uses organic molecules as terminal electron acceptors. Fermentation regenerates the supply of NAD$^+$ for glycolysis and results in the consumption of pyruvate and the release of molecules such as CO_2 or H_2 (gases); lactic, formic, acetic, succinic, butyric, or propionic acids; and ethanol, butanol, or propanol (alcohols). The final product depends on the species. No additional ATP is generated during fermentation.

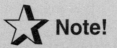 **Note!**

Many of the waste products of fermentation are valuable commercial products!

Respiration involves the oxidation of molecules, the generation of high-energy molecules, such as ATP, by passing pairs of electrons (and hydrogen ions, or protons) through an electron transport system, and the donation of these electrons to an inorganic electron acceptor. If the terminal electron acceptor is oxygen, this process is termed **aerobic respiration**. **Anaerobic respiration** occurs when the terminal electron acceptor is an inorganic molecule other than molecular oxygen (such as sulfate or nitrate). Organisms vary in their oxygen requirements; some are **strict anaerobes** and cannot survive in the presence of oxygen. **Facultative anaerobes** can respire aerobically or anaerobically, and **obligate aerobes** require oxygen for survival.

Pyruvate generated from glycolysis in the cytosol may enter the mitochondria and, if oxygen is available, be enzymatically converted to acetyl coenzyme A (acetyl CoA) and CO_2. Within the matrix of the mitochondria or the cytosol of aerobic prokaryotes, the two-carbon acetyl CoA

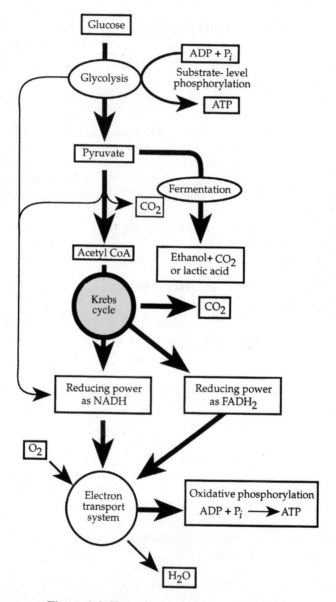

Figure 1-3 Chemoheterotrophic metabolism.

enters a circular set of enzymatic reactions known as the **Krebs cycle**, the **tricarboxylic acid cycle** (TCA), or the **citric acid cycle** (see Figure 1-3). During oxidation of a substrate, two major electron carriers, NAD^+ and FAD, become reduced to NADH and $FADH_2$. One complete turn of the TCA produces three molecules of NADH, two molecules of CO_2, one molecule of $FADH_2$, and one molecule of **guanosine triphosphate** (GTP). The electrons and H^+ ions from NADH and $FADH_2$ are transferred to the electron transport chains within the cristae of the mitochondria or the plasma membrane of prokaryotes. These chains consist of series of proteins that first serve as electron acceptors, then donors to the next complex in the chain. This series of coupled oxidations and reductions results in the terminal tranfer of electrons and H^+s to oxygen, forming water as the end product.

The complete oxidation of glucose:
$$C_6H_{12}O_6 + 6O_2 \rightarrow 6CO_2 + 6H_2O$$

ATP can be generated by three different mechanisms. It can be formed from adenosine diphosphate (ADP) by either **substrate-level phosphorylation** or **oxidative phosphorylation**. In substrate-level phosphorylation, an enzyme mediates the transfer of a phosphate group from a phosphorylated organic molecule to ADP. Oxidative phosphorylation occurs when molecules are oxidized and energy is extracted from the electrons by passing them through an electron transport system, where most of the resulting free enrgy is used to drive the phosphorylation of ADP, producing ATP. **Photophosphorylation** also synthesizes ATP, but uses the energy from sunlight rather than from the breakdown of organic molecules.

Reproduction

Most cells reproduce **asexually**, without exchanging or acquiring new hereditary information. Bacteria reproduce almost exclusively in this fashion in a process called **binary fission**, during which the bacterium grows, duplicates its hereditary information, segregates the duplicated chromosome, and divides the cytoplasm. Most cells that form the bodies of multicellular eukaryotes are also produced asexually in a process termed **mitosis**. During mitotic division, the cells grow, duplicate their genomes, separate their duplicated chromosome sets into nuclei at the opposite poles of the cell, and divide the cytoplasm to form progeny cells.

The eukaryotic **cell cycle** contains four major phases (see Figure 1-4). The **S phase** is when DNA synthesis occurs to replicate the chromosomes by creating identical **sister chromatids**. The period between S phase and the beginning of mitosis (**M phase**) is a gap, or growth period, designated **G_2 phase**. Another gap or growth period called the **G_1 phase**, occurs between the M and S phases to complete the cycle.

Mitosis consists of four consecutive phases: **prophase**, **metaphase**, **anaphase**, and **telophase** (see Figure 1-5). During prophase, each chromosome shortens and thickens by supercoiling on itself again and again.

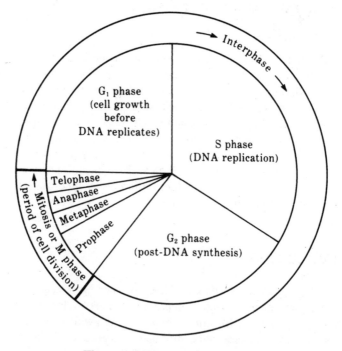

Figure 1-4 Eukaryotic cell cycle.

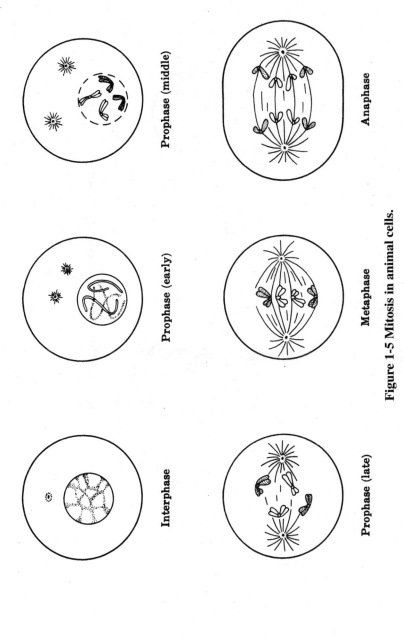

Prophase (middle)

Prophase (early)

Interphase

Anaphase

Metaphase

Prophase (late)

Figure 1-5 Mitosis in animal cells.

11

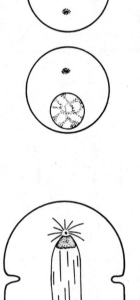

Telophase   **Daughter cells**

Figure 1-5 Mitosis in animal cells, continued.

12

The nuclear membrane dissolves, and a **spindle** of microtubules forms from one pole of the cell to the other. During metaphase, the chromosomes line up in the center of the spindle. At anaphase, the two chromatids of each replicated chromosome are pulled to opposite poles by depolymerization of the microtubules in the spindle apparatus that are attached to the centromeres. These former sister chromatids are now considered to be new chromosomes. Division of the cytoplasm (**cytokinesis**) begins in telophase, as the chromosomes unwind and new nuclear membranes form to enclose the sets of chromosomes at each pole of the cell. When mitosis is completed, two progeny cells contain identical sets of chromosomes.

The **somatic** cells of most plants and animals are **diploid**, meaning they have two sets of **homologous** chromosomes. One set is derived from each parent through the gametes that produced the zygote from which the organism developed. The process of **meiosis** reduces the chromosome number from diploid to **haploid** in gametes, or sex cells; thus, each parent contributes an equal number of chromosomes to their offspring.

You Need to Know ✔

Meiosis I is **reductional division,** since the number of chromosomes is reduced; meiosis II is **equational division.**

The predominant form of reproduction in most multicellular eukaryotes is **sexual.** At sexual maturity, some diploid germ line cells become specialized to undergo meiosis and form haploid gametes. Meiosis can be visualized as two highly modified cell cycles, back to back (see Figure 1-6). A complete meiotic cycle involves one initial DNA replication and two cytoplasmic divisions, yielding four haploid products, none of which are genetically identical. The two cycles are labeled meiosis I and II, each of which has its own prophase, metaphase, anaphase, and telophase.

The major events of these phases mirrors the events during mitosis. However, during prophase I of meiosis, homologous chromosomes pair

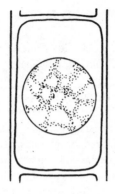

Interphase

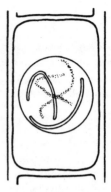

Early Prophase I

Synapsis

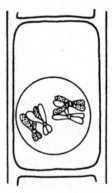

Crossing Over

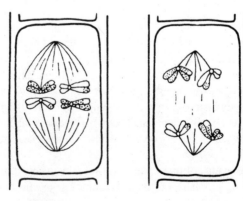

Metaphase I

Anaphase I

Figure 1-6 Meiosis in plant cells.

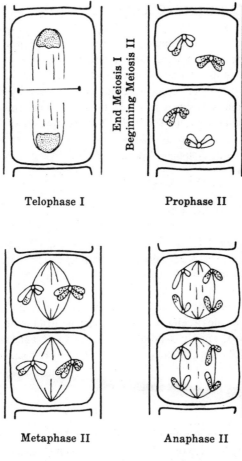

End Meiosis I
Beginning Meiosis II

Telophase I   Prophase II

Metaphase II   Anaphase II

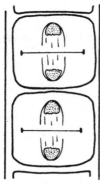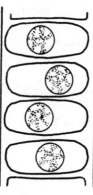

Telophase II   Meiotic Products

Figure 1-6 Meiosis in plant cells, continued.

up in a process called **synapsis**. A synapsed pair of chromosomes contains four chromatids. Each chromosome usually has one or more regions in which two of the four chromatids break at corresponding sites and reunite with one another, a process called **crossing over**, which increases genetic variability. During anaphase I, the homologous chromosomes are separated, yielding two haploid cells at the completion of the first stage of meiosis. During anaphase II, sister chromatids are separated, as they are during mitotic anaphase. The end result is four genetically different haploid cells.

Solved Problems

Solved Problem 1.1 Aside from DNA and certain associated proteins in chromosomes, what macromolecular aggregates are shared by all prokaryotes and eukaryotes?

Both prokaryotic and eukaryotic cells possess a lipid plasma membrane that separates a cell from its environment. In addition, all cells have ribosomes, made partly of protein and partly of ribonucleic acid (RNA) molecules. Ribosomes function in the synthesis of proteins.

Solved Problem 1.2 How are chloroplasts and mitochondria structurally similar?

They both are surrounded by an inner and outer membrane, a means for increasing the area of their membrane systems, contain their own circular chromosome, and have their own ribosomes.

Solved Problem 1.3 Why can't H_2S or NH_3 act as terminal electron acceptors in anaerobic respiration?

H_2S and NH_3 are both already completely reduced.

Solved Problem 1.4 What would you expect to happen if a facultative anaerobe were grown in the presence of oxygen and glucose?

If oxygen is present for aerobic respiration, fermentation essentially ceases, the rate of glucose consumption decreases, and the rate of acid and/or alcohol production is inhibited. This phenomenon is known as the **Pasteur effect**.

Solved Problem 1.5 What occurs in meiosis, but not mitosis?

Synapsis, crossing over, and separation of homologous chromosomes happen during meiosis, but not mitosis.

Chapter 2
BIOMOLECULES

Carbohydrates

Pure **carbohydrates** have the empirical formula $(CH_2O)_n$. The smallest carbohydrates are simple sugars, or **monosaccharides**. Glucose is the six-carbon monosaccharide (**hexose**) used as a basic source of energy by most heterotrophic cells. Ri-

bose and deoxyribose are the five-carbon sugars (**pentoses**) that serve a structural role in the nucleic acids RNA and DNA, respectively. **Oligo-saccharides** are small polymers of two to six monosaccharides. **Sucrose** is a disaccharide of the two monosaccharides glucose and **fructose** (an isomer of glucose). Sucrose is the major sugar transported between plant cells, whereas glucose is the primary sugar transported between animal cells. **Lactose**, the major sugar in milk, is a disaccharide of glucose and **galactose** (an epimer of glucose). Most of the carbohydrate molecules in

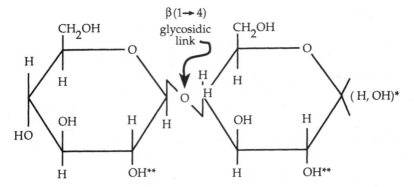

Figure 2-1 Cellobiose, the basic repeating unit of cellulose, is a disaccharide of glucose molecules joined by $\beta(1\rightarrow4)$ glycosidic linkages.

nature are composed of hundreds of sugar units and are referred to as **polysaccharides**.

The monomers of polysaccharides become covalently connected by **glycosidic bonds** (see Figure 2-1).

Carbohydrates serve several major functions in living systems. Monosaccharides and oligosaccharides serve as readily utilizable energy sources. Starch and glycogen act as macromolecular energy stores in plants and animals, respectively. Carbohydrates perform structural roles, such as cellulose in plant cell walls and chitin in the exoskeletons of arthropods. Surface carbohydrates are often complexed with proteins as **glycoproteins** or with lipids as **glycolipids** in the plasma membrane. The great potential for structural diversity and thus, specificity, makes these molecules very useful as cell-recognition markers in cellular communication and in cell-to-cell attachments.

 **Note!**

Glycogen consists of polymers of glucose units joined by $\alpha(1\rightarrow4)$ linkages and forms branched chains by $\alpha(1\rightarrow6)$ linkages. Starch has fewer $\alpha(1\rightarrow6)$ linkages than glycogen.

Lipids

Lipids are water-insoluble (**nonpolar**) molecules that are soluble in weakly polar or nonpolar solvents such as chloroform. The most important function that lipids perform for all kinds of cells stems from their ability to form sheetlike membranes. The plasma membrane of both prokaryotic and eukaryotic cells separates the cellular contents from the external environment, thus allowing the cell to function as a unit of life. Eukaryotic cells also have internal membranes, such as those of the ER, nucleus, mitochondrion, and chloroplast, that further compartmentalize the cell for specific functions. The other important function of lipids is as efficient energy storage molecules.

There are three major kinds of membrane lipids: phospholipids, glycolipids, and sterols. Both phospholipids and glycolipids readily associate spontaneously to form a **lipid bilayer** (see Figure 2-2). Cellular membranes behave as two-dimensional, semifluid structures, allowing embedded protein molecules to constantly move about rather freely by lateral diffusion. The fluidity of prokaryotic membranes is regulated by varying the number of double bonds in, and the lengths of, the fatty acid chains of the lipid molecules constituting the membrane. In animals, the quantity of the sterol lipid cholesterol is a key regulator of membrane fluidity.

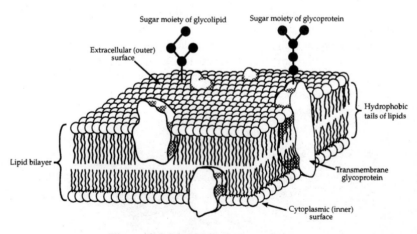

Figure 2-2 Lipid bilayer membrane.

The plasma membrane is a selective filter that controls the entry of nutrients and other molecules needed for cellular processes. Waste products of metabolism pass out of the cell through this membrane. Due to their composition, membranes have a low permeability for ions and most polar molecules, thus these molecules must pass through channels formed from integral membrane proteins. If a substance is moving against its concentration gradient (i.e., from an area of lower concentration to an area of higher concentration), then energy must be expended. This is termed **active transport**.

Proteins

Proteins consist of chains of 20 different kinds of amino acids connected by covalent linkages called **peptide bonds**. All amino acids have the same generalized structure as shown in Figure 2-3. An **α-carbon** is at the center of each amino acid. To its left (as conventionally written) is a basic (when ionized) amino group (NH_3^+). To the right of the α-carbon is an acidic (when ionized) carboxyl group (COO^-). A hydrogen atom forms a third bond to the α-carbon, and the fourth bond connects to a side-chain group (R).

Amino acids are classified according to the nature of the R group. The 20 different amino acids used in the synthesis of proteins are symbolized by either three letter or single letter abbreviations as listed in Table 2.1.

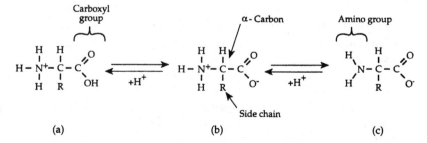

Figure 2-3 Generalized structure of amino acids at different pH values. Predominant forms in (a) acidic, (b) neutral (pH 7), and (c) basic solutions.

Type	Amino Acid	Abbreviation	
Nonpolar, Aliphatic	Glycine	Gly	G
	Alanine	Ala	A
	Valine	Val	V
	Leucine	Leu	L
	Isoleucine	Ile	I
Polar, Aliphatic	Serine	Ser	S
	Threonine	Thr	T
	Asparagine	Asn	N
	Glutamine	Gln	Q
Aromatic	Phenylalanine	Phe	F
	Tyrosine	Tyr	Y
	Tryptophan	Trp	W
Sulfur-Containing	Cysteine	Cys	C
	Methionine	Met	M
With Secondary Amino Group	Proline	Pro	P
Acidic	Aspartate	Asp	D
	Glutamate	Glu	E
Basic	Lysine	Lys	K
	Arginine	Arg	R
	Histidine	His	H

Table 2.1 Amino acids grouped by chemical type.

Peptide bonds linking amino acids are enzymatically formed by dehydration synthesis. An oxygen atom is removed from the carboxyl group of one amino acid together with two hydrogens from the amino group of a second amino acid (see Figure 2-4). This gives peptide chains polarity. At one end is a free amino group, at the other, a free carboxyl group. **Oligopeptides** are chains usually less than ten amino acids in length. The term **polypeptide** is used for longer chains of amino acids, whereas chains over 5,000 daltons are generally called **proteins**. Some proteins consist of only a single polypeptide chain. In these cases, a complete polypeptide chain would be synonymous with a functional protein. In other instances, however, a functional protein may consist of two or more chains.

You Need to Know ✔

An average polypeptide contains about 300 residues.

Polypeptides may differ by the number and kinds of individual amino acids they contain. The final structure can be described on four levels of increasing complexity. The **primary structure** of a functional protein consists of the linear sequence of amino acids in each of its polypeptide chains. There are two major kinds of **secondary** protein structure: **α-helix** and **β-pleated sheet**. An α-helix forms when a carbonyl (C=O) adjacent to one peptide bond is linked by a hydrogen bond to an amino group (NH) flanking a peptide bond in an amino acid about

four residues along the same chain. β-pleated sheets form when hydrogen bonds form between amino acids on adjacent, parallel polypeptide strands. The polypeptide chain may fold back upon itself, forming weak, internal bonds (e.g., hydrogen bonds, ionic bonds) as well as stronger covalent disulfude bonds that stabilize its **tertiary structure** into a precisely and often intricately folded pattern. These bonds are formed from the side chains of different amino acid residues. If two or more polypeptide chains spontaneously associate, they form a **quaternary structure**.

Proteins perform many enzymatic, structural, and other roles in living systems. For example, they are a major structural component of ribosomes, they may act as hormones that signal between different cell

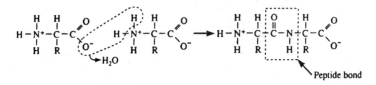

Figure 2-4 Dehydration synthesis of a dipeptide by the formation of a peptide bond.

types, or they may assist in the movement of organelles within the cell and movement of the cell itself.

Nucleic Acids

Nucleic acids occur in two forms, **deoxyribonucleic acid** (DNA) and **ribonucleic acid** (RNA). Both are linear, unbranched polymers of subunits termed **nucleotides**. DNA is found in the nucleus of eukaryotes and the cytoplasm or **nucleoid** of prokaryotes and functions as the molecule of heredity (see Chapter 3). RNA molecules are synthesized on DNA templates and participate in protein synthesis in the cytoplasm (see Chapters 4 and 5).

Each nucleotide consists of three major parts: (1) a five-carbon sugar (pentose); (2) a flat, heterocyclic, nitrogen-containing organic base; and (3) a negatively charged phosphate group, which gives the polymer its acidic property (see Figure 2-5). The nitrogenous base in each nucleotide is covalently attached to the sugar by a glycosidic bond. The phosphate group is also covalently linked to the sugar.

The sugar β-D-ribose is found in ribonucleotide monomers of RNA. The pentose in the deoxyribonucleotide monomers of DNA differ by the absence of oxygen at the #2 carbon and is thus 2-deoxy-β-D-ribose.

The organic bases are of two general types: single-ringed **pyrimidines** and double-ringed **purines**. The purines are **adenine** (A) and **guanine** (G). The pyrimidines are **cytosine** (C), **thymine** (T), and **uracil** (U). Thymine is found primarily in DNA and uracil is found only in RNA. In each polynucleotide strand of DNA and RNA, adjacent nucleotides are joined covalently by **phosphodiester bonds** between the 3' carbon of one nucleotide and the 5' carbon of the adjacent nucleotide.

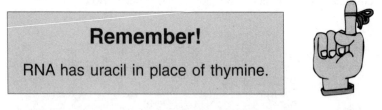

Remember!

RNA has uracil in place of thymine.

Bases in the nucleotides spontaneously form hydrogen bonds in a highly specific manner. Adenine normally forms two hydrogen bonds with thymine in a **complementary** strand of the DNA double helix. Like-

The Sugars

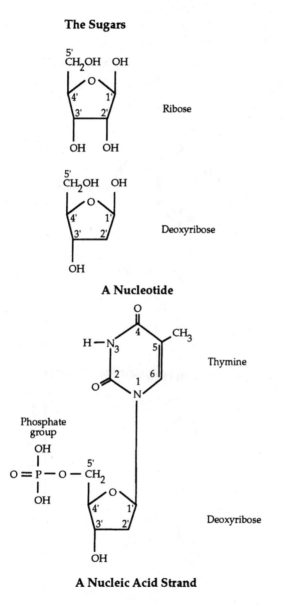

A Nucleotide

A Nucleic Acid Strand

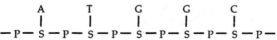

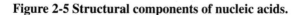

Figure 2-5 Structural components of nucleic acids.

The Bases

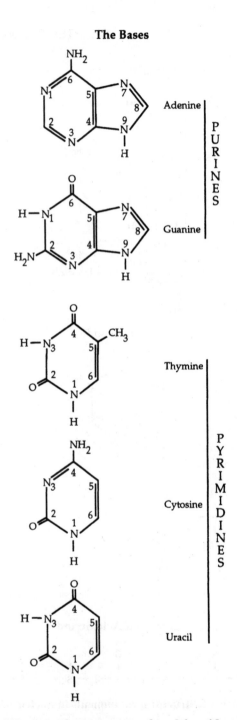

Figure 2-5 Structural components of nucleic acids, continued.

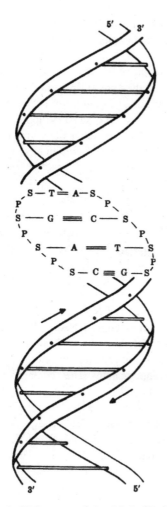

Figure 2-6 Diagram of double helical DNA.

wise, it can form two hydrogen bonds with U in DNA-RNA hybrids and in RNA-RNA interactions. Guanine forms three hydrogen bonds with cytosine. DNA exists in the uniform shape of a double helix (see Figure 2-6), with the complementary chains wound around each other like a spiral staircase, whereas RNA molecules are synthesized from DNA templates as single strands. The single strand of RNA, however, may fold back onto itself and form complementary base pairs to make unique secondary structures.

The two complementary strands of a DNA double helix run in opposite directions, that is, they are **antiparallel**. If one chain is read from the 5′ phosphate end, the other would read from the 3′ hydroxyl. The double helix makes a turn every ten base pairs (approximately 3.4 nm). The paired bases are stacked in the center of the molecule, forming a hydrophobic core and giving the helix a width of about 2 nm.

 **Note!**

Since A always pairs with T, and G always pairs with C, the purine:pyrimidine ratio in double stranded DNA is always 1.

There are three classes of RNA based on their functions: (1) **transfer** RNAs (tRNAs); (2) **messenger** RNAs (mRNAs); and (3) **ribosomal RNAs** (rRNAs). The tRNAs are the smallest (75-80 nucleotides in length) and serve to position each amino acid on the ribosome for polymerization into polypeptide chains. They contain a few unusual bases in addition to A, C, G, and U. The genetic code that specifies the amino acid sequences of proteins resides in the DNA sequence, and it becomes transcribed into complementary ribonucleotide sequences of mRNA, thus the length and composition of different mRNAs can vary greatly. The rRNAs are structural components of the ribosomes. There are three classes of rRNAs in bacteria and four in eukaryotes.

Solved Problems

Solved Problem 2.1 What is the composition of starch? How is it digested?

Starch is a homopolymer of glucose units joined in $\alpha(1\rightarrow4)$ and $\alpha(1\rightarrow6)$ linkages. During digestion by enzymes such as salivary and pancreatic amylases, starch is hydrolyzed to maltose and glucose. Maltose is a disaccharide of two glucoses units joined by an $\alpha(1{\to}4)$ link that can be cleaved by the enzyme maltase to yield two glucose molecules.

Solved Problem 2.2 Would you expect certain amino acids to have a preferential location within a protein?

The ionized side chains of some amino acids readily interact with water (hydrophilic). Hydrophobic amino acids contain nonionized side chains that prefer to avoid contact with water. Thus, when a polypeptide chain folds into a globular tertiary shape, amino acids with hydrophilic groups tend to predominate on the outside of the molecule and hydrophobic segments of the chain tend to predominate in the interior of the molecule.

Solved Problem 2.3 How do RNA molecules structurally differ from DNA molecules?

RNA contains uracil rather than thymine, has ribose rather than deoxyribose as the pentose sugar, and is usually single-stranded, whereas DNA is usually double stranded.

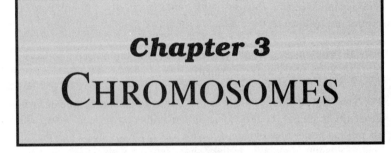

Chapter 3
CHROMOSOMES

Chromosome Structure

All essential bacterial genes are found in a single, circular, double-stranded DNA **chromosome** located in the nucleoid region of the cytoplasm. The bacterial chromosome is believed to be attached to the plasma membrane and specifies between 1,000 and 5,000 proteins. It is highly condensed and consists of DNA, RNA, and protein. In addition, there may be one or more **plasmids**. Plasmids are small circular pieces of extrachromosomal DNA which may encode 20–100 proteins.

The genes of eukaryotes are distributed among a number of linear chromosomes that vary in size and number. Eukaryotic chromosomes are condensed by packing the DNA to different degrees (see Figure 3-1). **Nucleosomes** consist of DNA wound twice around an octet of proteins called **histones** (two each of **H2a**, **H2b**, **H3**, and **H4**). Approximately 200 base pairs (bp) of the DNA are wound around the spherical bodies formed by the histones, and about 50 bp of

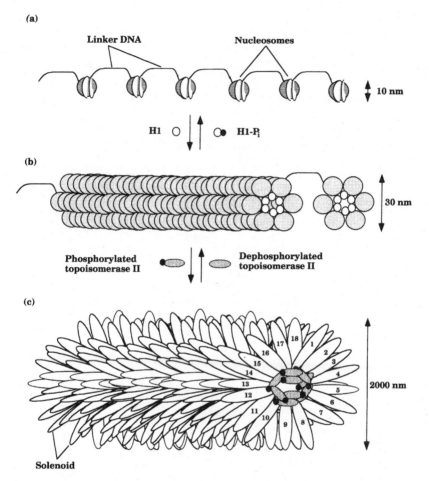

(a)

Linker DNA   Nucleosomes

10 nm

H1   H1-P$_i$

(b)

30 nm

Phosphorylated topoisomerase II   Dephosphorylated topoisomerase II

(c)

2000 nm

Solenoid

Figure 3-1 Eukaryotic chromosome packaging: (a) extended nucleosome form; (b) solenoid form; and (c) looped solenoid form.

DNA connect the nucleosomes. Further compaction may be accomplished by histone **H1** binding, which induces the nucleosomes to associate into a ring of six nucleosomes and the rings to associate into a cylinder called a solenoid. Phosphorylation of histone H1 results in the dissociation of the solenoid into an extended nucleosome form. The solenoid is the form in which most of the cell's DNA exists during inter-phase. However, further packing can occur by certain proteins binding the solenoid and stimulating it to loop back and forth from a central core of proteins called a **scaffold**. Dephosphorylation of **topoisomerase II** and other proteins causes dissociation of the scaffold and results in the decondensation of the chromosomes to the solenoid form. In some eu-karyotes, 18 loops of the solenoid form a disklike structure and the chro-mosome condenses as hundreds of disks stack together. This is the form that is predominant during nuclear division.

Let's Compare!

Yeasts have 4 chromosomes; haploid human cells have 23.

Heterochromatin is highly condensed DNA that remains in the so-lenoid form throughout the cell cyle except during DNA replication, when it decondenses. Most of the genes associated with heterochromatin are not expressed because of the DNA's condensed state. In contrast, **eu-chromatin** is decondensed DNA that exists in the solenoid form or in an extended nucleosome form.

Remember

Euchromatin in the nucleosome form can be expressed; in the solenoid form it cannot.

A **centromere** is a highly constricted region of a mitotic or meiotic chromosome where the spindle fibers attach. Complex sequences of DNA constitute centromeres. If the centromere is in the middle of the chromosome, the chromosome is said to be **metacentric**. If the cen-tromere is near the tip, it is called **telocentric**. The short and long arms of the chromosome with respect to the centromere are designated as **p** and **q**, respectively. Special staining techniques reveal that each chromosome has a specific pattern of dark and light regions called bands. Homologous chromosomes have the same banding pattern.

Protein complexes associated with the centromeric regions are called **kinetochores**. Kinetochores bind microtubules of the spindle bundle and function to distribute chromosomes as cells proliferate.

Propagation and maintenance of any piece of DNA requires the presence of one or more **origin of replication sites** (*OriR*) and special ends called **telomeres**. Origin of replication sites are special sequences where DNA replication initiates. Telomeres protect the ends of linear chromosomes from cellular enzymes that degrade nucleic acids from their ends.

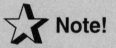 **Note!**

Bacterial genomes range from 10^6 to 3×10^7 bp, whereas a diploid human cell has 5.6×10^9 bp among its 46 chromosomes. However, 90–95% of the human genome is not expressed as protein.

DNA Replication

DNA replication of the *E. coli* chromosome begins at a single origin of replication (*oriC*) and proceeds bidirectionally to a termination site located approximately halfway around the circular chromosome. During replication, the DNA strands of the double helix must be both unwound and separated.

DNA replication is initiated when a protein encoded by the gene ***dnaA*** binds repetitive 9-mer sequences at the origin. Subsequently, **helicases** specified by ***dnaB*** and inhibitory proteins encoded by ***dnaC*** bind repetitive 13-mer sequences. As helicase progresses 5' to 3', dissociation of protein DnaC allows the helicase to unwind the DNA. The unwinding produces positive superhelical turns in the rest of the DNA, making it energetically favorable to continue unwinding the strands. To unwind the DNA, positive superhelical turns have to be removed by cutting the DNA and allowing it to relax or by introducing negative superhelical turns to compensate for the positive ones. The introduction of negative superhelical turns requires energy and an enzyme called **DNA gyrase** (a topoisomerase). DNA gyrase is an enzyme that can both remove positive supercoils or introduce negative supercoils into the DNA and thereby make strand separation energetically more favorable. Presumably the DNA gyrase binds ahead of the unwound DNA during replication. **Single-stranded binding proteins** (SSBPs) act to temporarily stabilize the unwound state.

DNA replication (see Figure 3-2) begins with the synthesis of a 30 nucleotide long RNA **primer** by an RNA polymerase called **primase** (specified by *dnaG*). The helicase and primase subsequently form a complex enzyme system known as the **primosome**, which synthesizes primers after DNA synthesis begins. Two catalytic subunits of **DNA polymerase III** (PolC) associate with the templates and the 3′ ends of the primers and begin to polymerize deoxyribonucleotides into DNA. DNA gyrase continues to remove positive supercoils and/or introduces negative supercoils ahead of the primosome that is opening the two strands of DNA. At various intervals, the template signals the primase portion of the primosome to polymerize primer RNAs about 30 nucleotides long on only one template at the replication fork. DNA polymerase III polymerizes DNA 5′ to 3′ from each of the primers at the replication fork. One strands of DNA is polymerized toward the replication fork and continues to be elongated as the DNA unwinds further. The second strand of DNA is polymerized away from the replication fork. As the DNA unwinds further, a new primer is synthesized away from the replication fork and the DNA polymerase synthesizes DNA from the last primer toward the previous RNA primer. As the DNA polymerase reads the template strand, it selects complementary nucleotides for the nascent strand based on hydrogen bonding capability.

The DNA synthesized toward the replication fork is synthesized in a continuous manner and is called the **leading** strand. The opposite DNA strand is synthesized in a discontinuous manner away from the replication fork and is referred to as the **lagging** strand. The leading and lagging strands are synthesized halfway around the bacterial chromosome until they encounter the lagging and leading strands synthesized at the other replication fork.

The RNA-DNA fragments that initially constitute the lagging strand are known as **Okazaki** fragments, named after the scientist who discovered them.

The RNA primers are removed by a DNA repair enzyme called **DNA polymerase I** specified by *polA*. It uses neighboring DNA as a primer

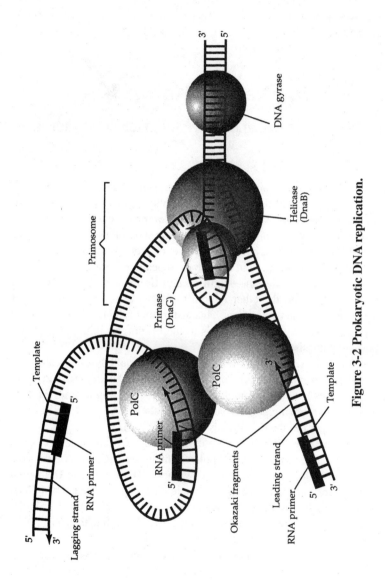

Figure 3-2 Prokaryotic DNA replication.

and polymerizes DNA from it, displacing the RNA primer. A DNA ligase removes nicks in the DNA by connecting the fragments together. **Topoisomerase IV** is required to separate the two daughter chromosomes.

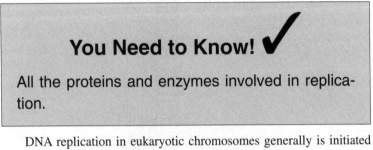

You Need to Know! ✔

All the proteins and enzymes involved in replication.

DNA replication in eukaryotic chromosomes generally is initiated from many origin of replication sites. Replication forks proceed in both directions from these sites. The sites that comprise yeast origins of replication are called **autonomously replicating sequences** (ARSs) and consist of two regions that bind a distinct set of proteins that destabilize the double helix. In one region, conserved, repeating 11-mers bind a multiprotein complex called the **origin recognition complex** (ORC). When proteins also bind the other region, the DNA bends by interaction of the proteins in the two regions. This distortion of the DNA promotes the separation of paired DNA strands at the origin and initiation of RNA primer synthesis.

Enzymes similar to those involved in bacterial DNA replication are found in eukaryotes. Numerous topoisomerases, helicases, and RNA polymerases have been found in eukaryotes. **DNA topoisomerase II** is involved in relieving positive supercoils in the DNA, whereas a helicase activity separates the two strands (see Figure 3-3).

At least five different DNA polymerases have been found in eukaryotic cells. The primase (**DNA polα**) synthesizes lagging strand DNA. **DNA polδ** catalyzes leading strand synthesis. **DNA polϵ** and **DNA polβ** are responsible for replacing the nucleotide gaps created when RNA primers are removed by endonucleases. A DNA ligase repairs single-stranded nicks (unconnected adjacent nucleotides) left in the DNA. **DNA polγ** performs DNA replication in the mitochondria.

To complete replication of a linear chromosome, RNA primers at each end of the chromosome have to be removed and replaced by DNA. Although RNA primers can be removed by exonucleases, none of the usual DNA polymerases are able to replace the RNA without a DNA

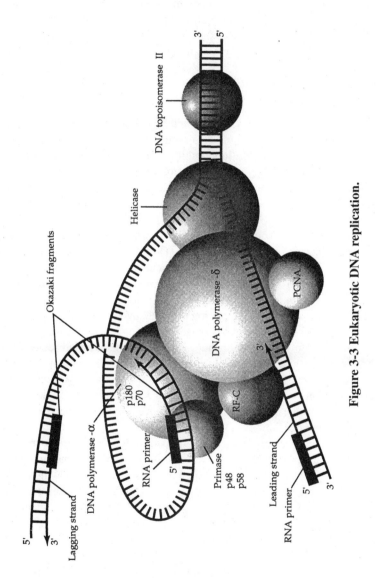

Figure 3-3 Eukaryotic DNA replication.

primer. An unusual type of DNA polymerase known as **telomerase** consists of protein and an RNA template that the protein portion copies repetitively into DNA in order to extend one strand of the telomere. Thus, telomerase is responsible for maintaining the length of the chromosomes.

Recombination

In bacteria, pieces of DNA that can enter a cell may become part of the main chromosome or one of its plasmids. The integration process is termed **genetic recombination**, and it generally occurs at points where the two DNAs are nearly identical. Recombination can be beneficial because it can create new genetic information. In addition, recombination provides a mechanism by which organisms can replace genes that have been severely damaged or even deleted.

In eukaryotes, recombination leads to genetic diversity among progeny produced by sexual reproduction. During meiosis, the process of crossing over produces linkage arrangements in gametes different from those that exist in the parent. Favorable gene combinations tend to be perpetuated by natural selection. The recombination that takes place during meiosis is also a means of repairing or replacing DNA.

There are two major types of genetic recombination. **Site-specific recombination** requires short but identical double-stranded regions of homology between recombining molecules of DNA, and it usually changes the relative positions of chromosomal segments. **General recombination** occurs between homologous DNA molecules. It does not normally alter the order in which gene loci occur in their respective chromosomes but does involve DNA synthesis.

In general recombination, if the transforming DNA is single-stranded, the **RecA** protein binds to it, then to the bacterial chromosome and melts it while searching for a region of homology. The minimal region of homology is about 60 bp but usually involves hundreds of bases. Hydrogen bonding occurs between the transforming DNA and a complementary region of the cell's DNA. An enzyme such as **UvrABC** cuts away the unpaired portion of the melted DNA. A ligase repairs the nicks. If there is a mismatch, UvrABC cuts on either side of a mismatched strand, and a repair enzyme (DNA polymerase I) pushes away the mismatched strand and replaces it with new DNA that matches the complementary strand. If the transforming DNA that enters the cell is double-stranded, the **RecBCD** protein searches the DNA for a specific sequence called a

chi (χ) **site**, where it cuts one strand of the DNA, creating a nick. The single stranded DNA that results is coated with RecA, which then searches the bacterial chromosome for a region of homology and the remaining steps are similar to those discussed above.

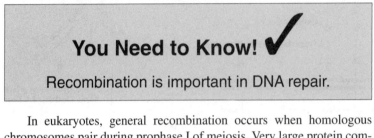

You Need to Know! ✔

Recombination is important in DNA repair.

In eukaryotes, general recombination occurs when homologous chromosomes pair during prophase I of meiosis. Very large protein complexes called **recombination modules** are found at intervals along the **synaptonemal complex**, the ladderlike protein structure that develops between chromosome pairs. At each recombination module, two of the four chromatids break and rejoin with one another in crossing over. It is believed that endonucleases in the recombination modules nick a single strand of each chromatid capable of recombination and that helicases unwind the DNA, creating single-stranded regions. A protein similar to RecA is proposed to catalyze the pairing of the single-stranded DNAs to the complementary strands on the homologous chromatids. A DNA polymerase may extend the exchanged strands, and a DNA ligase is thought to eliminate the nicks in the strands. This model is referred to as the **Holliday** model after the geneticist who proposed it (see Figure 3-4).

Solved Problems

Solved Problem 3.1 How does DNA replication differ between eukaryotes and prokaryotes?

The basic mechanisms of prokaryotic and eukaryotic DNA replication are similar. However, eukaryotes have numerous linear chromosomes, each with many *oriR* sites, whereas prokaryotes have only one replication origin on a single circular chromosome. Eukaryotes have more DNA polymerases than prokaryotes. Different proteins bind to the origins to initiate unwinding of the DNA prior to replication.

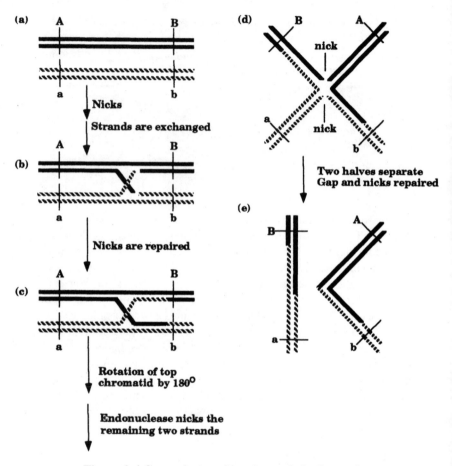

Figure 3-4 General recombination and the formation of a Holliday intermediate.

Solved Problem 3.2 If the GC content of a DNA molecule is 48%, what are the percentages of each of the four nucleotides?

Since each G hydrogen bonds with a C on the complementary strand of double-stranded DNA, then G=C. Thus, if the GC content is 48%, the G content must be 24% and the C content must be 24%. This leaves 52% for A and T, which must be 26% each.

Chapter 4
TRANSCRIPTION AND GENE REGULATION

IN THIS CHAPTER:

- ✔ Introduction
- ✔ Prokaryotic Genes
- ✔ Transcription Initiation and Termination
- ✔ The Lac Operon
- ✔ Eukaryotic Gene Regulation
- ✔ RNA Processing
- ✔ Solved Problems

Introduction

The central dogma of molecular genetics proposes that the information in DNA is used to make RNA molecules through a process known as **transcription** and that the information in some RNA is used to make proteins by a process called **translation**. Transcription is carried out by RNA polymerases, whereas translation is catalyzed by enzymes associated with ribosomes. The RNA molecules and proteins synthesized during the development and/or maintenance of an organism are responsible for an organism's characteristics.

41

Remember!

Gene expression includes both transcription and translation!

The information for synthesizing a particular RNA is located in only one of the two DNA strands. The strand that contains the information for making an RNA molecule and that is "read" by an RNA polymerase is called the **template** strand or the **sense** strand. The strand that is complementary to the template is sometimes called the **nonsense** strand since it provides no information for the making of RNA or protein. Not all templates for coding RNAs occur on the same strand of DNA, however. Messenger RNA that specifies the synthesis of a protein is called **sense RNA**, whereas RNA complementary to sense RNA is called **antisense RNA**.

Most genes, especially those encoding proteins, are regulated so they are expressed at the appropriate time and level needed to maintain the cell or to promote its growth and proliferation.

Prokaryotic Genes

Structural genes are the nucleotide sequences of DNA that serve as templates for the synthesis of RNAs. The average length of structural genes specifying proteins in prokaryotes is about 1,000 bp, compared to about 10,000 bp in eukaryotes. Structural genes and the controlling sites that regulate the rate of transcription of these genes are called **operons** (see Figure 4-1).

Regulatory proteins are proteins that affect the expression of structural genes by binding to controlling sites near the structural genes and either activating or repressing transcription. Regulatory proteins that stimulate gene transcription are termed **transcription factors** or **activators**. **Repressors** are proteins that inhibit the initiation of transcription when bound to controlling sequences called **operators**. Proteins that terminate transcription are referred to as **terminator proteins**. However, the end of transcription is most often signaled by a specific **terminator** sequence found in the DNA or newly transcribed RNA. In general, activa-

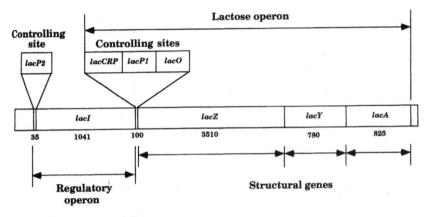

Figure 4-1 Structure of the lactose operon.

tors stimulate RNA polymerase binding to **promoter** sites on DNA at the beginning of structural genes, whereas repressors inhibit RNA polymerase binding. **Controlling sites** are short nucleotide sequences of DNA, usually 15–30 bp long, that control the expression of structural genes next to them.

> ✴ **Know the controlling sites!**
>
> Promoters
> Operators
> Initiators
> Attenuators
> Terminators

The simplest operon consists of one structural gene and one promoter to serve as a binding site for RNA polymerase. These operons are **constitutive**, that is, they are expressed at all times. The regulatory operon in Figure 4-1 is an example of a simple operon. Some simple operons may be regulated by an **attenuator**, which codes for RNA structure that causes the RNA polymerase to prematurely cease transcribing.

Most operons in bacteria consist of numerous structural genes and controlling sites. In bacteria, many operons contain more than one structural gene, or **cistron**. These **polycistronic operons** are transcribed into a single mRNA. Each protein-coding region in the mRNA is defined by its own **start codon**, where protein synthesis is initiated, and a **nonsense codon**, where protein synthesis is terminated. Operons in eukaryotes are usually monocistronic, that is, they contain a single gene. A **regulon** is a group of operons under the control of a regulatory protein. The operons in a regulon are generally not contiguous.

Transcription Initiation and Termination

Transcription of a structural gene can only occur if there is a promoter for RNA polymerase binding near the beginning of the gene. In bacteria, promoters are usually 15–30 bp long. The base pair sequence of the promoter determines the efficiency with which RNA polymerase binds to it and thus the efficiency of transcription. In many cases, RNA polymerase binding to bacterial promoter sites is dependent upon a family of proteins called **sigma (σ) factors**. Each type of sigma factor determines the kind of promoter that RNA polymerase will recognize. Both a sigma factor and RNA polymerase are required for efficient binding to a promoter. After initiation, the sigma factor will dissociate from the RNA polymerase (see Figure 4-2a–c).

Although prokaryotic promoters vary considerably, two short regions are shared in common by most promoters. A region on the nonsense strand about 10 bp before transcription initiates (−10 sequence) has a consensus 5′-TATAAT-3′ and another region about 35 bp up (−35 sequence) has a consensus sequence 5′-TTGACA-3′. A **consensus sequence** contains the nucleotide sequence most commonly encountered in a wide variety of promoters.

Most terminators at the end of a mRNA molecule code for a single-stranded region that folds on itself due to hydrogen bonding between complementary base pairs and codes for a final region containing many uracils. The hairpin structure that forms interacts with the RNA polymerase and stimulates its detachment from the DNA. Some terminators require **terminator proteins** such as **Rho** (ρ) to be active (see Figure 4-2d,e). These Rho-dependent terminators lack a poly-U region in the 3′ end of the terminator. The **NusA** protein is another termination factor that

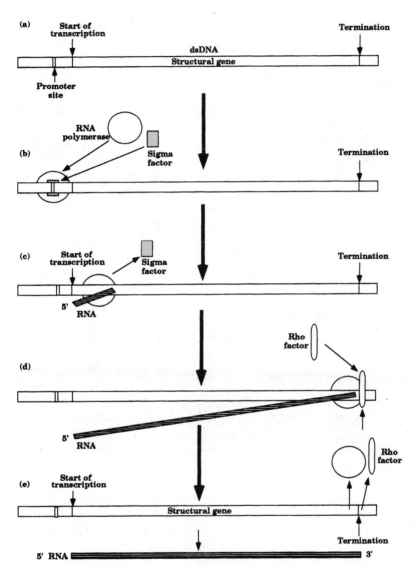

Figure 4-2 Roles of Sigma and Rho.

apparently binds directly to the RNA polymerase and catalyzes its release when it comes to termination sites.

The Lac Operon

The lactose operon (see Figure 4-1) provides a good model system of several concepts of prokaryotic gene regulation. It consists of three structural genes (*lacZ*, *lacY*, and *lacA*) as well as three controlling sites (*lacCRP*, *lacP1*, and *lacO*). The structural genes *lacZ*, *lacY*, and *lacA* encode the enzymes **β-galactosidase**, **permease**, and **transacetylase**, respectively. Catabolism of lactose is dependent upon these proteins. The controlling sites *lacCRP*, *lacP1*, and *lacO* are binding sites for the cAMP receptor protein, RNA polymerase, and the **lactose repressor**, respectively.

The lactose regulatory operon consists of one structural gene (*lacI*) and one controlling site (*lacP2*). The structural gene encodes the lactose repressor, whereas the controlling site is an RNA polymerase binding site (a promoter).

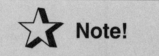

★ Note!

Because the lactose operon is *repressed* by a regulatory protein, it is said to be **negatively controlled**. Other operons are **positively** controlled, that is regulated by proteins that *activate* the operon.

In the absence of **inducer**, something that turns on transcription of an operon, expression is inhibited by the binding of the lactose repressor at *lacO*. The repressor sterically hinders the binding of RNA polymerase to *lacP1* and the inititaion of transcription. If lactose is present, it can be converted by the cell into **allolactose**, which acts as an inducer for this operon. When inducer is present, it binds to the repressor (protein LacI) and inactivates it. Inactive LacI cannot bind the operator, and RNA polymerase is able to bind to *lacP1* and initiate transcription of the genes necessary for lactose catabolism.

Even when an operon is induced the cell does not rapidly fill with mRNA and protein since mRNA has an average half-life of only 2.5 minutes. That means that 2.5 minutes after mRNA is synthesized, half of it will be degraded. Proteins are more stable than mRNA. Also, as the cell synthesizes mRNA and proteins, it depletes its stores of certain energy molecules. When a cell is rapidly metabolizing, **catabolite repression** shuts down many catabolic operons, including the lactose operon. This involves a small molecule called **cyclic adenosine monophosphate** (cAMP). The cellular level of cAMP decreases when the synthesis of lactose mRNA and enzymes increases, and the level of cAMP increases when these catabolic genes are no longer expressed. The higher the cAMP level, the more cAMP binds to a protein called **cyclic AMP receptor protein** (CRP), which then undergoes a conformational change that promotes binding to an activator binding site (*lacCRP*). This in turn upregulates transcription of the lactose operon genes.

Remember!

Lactose present → operon on
Lactose absent → operon off

Eukaryotic Gene Regulation

In eukaryotes, just as in bacteria, genes are regulated so that they are expressed at the right time and at the correct levels to maintain the cell or promote growth and proliferation. Cells in multicellular organisms must not only respond to the chemicals in their environment but must also work together through complex signaling that often affects gene activity.

The controlling sites in eukaryotes are similar to those found in bacteria, however there are many more controlling sites and proteins affecting each eukaryotic gene. Transcription factors (TFs) attach to binding sites in promoter regions and stimulate RNA polymerase binding to the promoter site (see Figure 4-3a,b). Transcription factors are produced constitutively since great numbers of genes depend on these for expression.

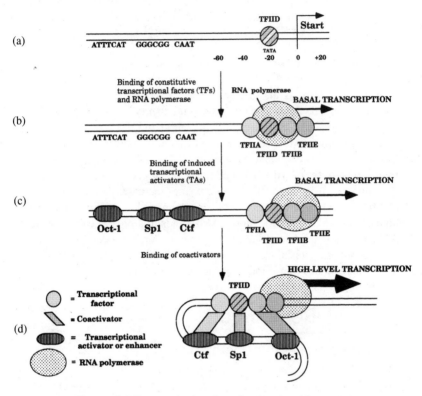

Figure 4-3 Transcriptional activation in eukaryotes.

Enhancers are bound by transcription activators (TAs) that are synthesized in response to specific signals. Most enhancers, such as the one that binds Gal4, are located hundreds or thousands of base pairs from the promoter sites. However, some induced activators, such as Fos and Jun, bind very near promoter sites. TAs cause the DNA to loop back on itself when they interact with the TFs near the promoter. This interaction between enhancer sites and initiator sites is usually necessary for transcription above a basal level (Figure 4-3c). **Coactivators** are activator proteins that often connect TFs and/or TAs with the RNA polymerase and may be essential for gene expression (Figure 4-3d).

Whereas only one RNA polymerase functions in bacterial cells, three different RNA polymerases are involved in eukaryotic nuclear transcription. The three polymerases initiate transcription only with specific combinations of TFs and TAs. **RNA polymerase I** transcribes the genes that specify 5.8S, 28S, and 18S rRNA. This polymerase is often found associated with chromosomes in the nucleoli. **RNA polymerase II** transcribes from promoters that control the synthesis of **pre-mRNA**, which consists of coding (**exons**) and noncoding (**introns**) regions. **RNA polymerase III** recognizes promoters that control the synthesis of relatively short RNAs such as tRNAs, 5S rRNA, and others.

In summary:

RNA pol I   →   rRNA
RNA pol II   →   mRNA
RNA pol III →   tRNA (and 5S rRNA)

The binding of RNA polymerase to promoter sites is dependent on a number of TFs such as the **TFIID complex** (see Figure 4-3a), which is functionally comparable to sigma factors in bacteria. TFIID is the first to bind close to the promoter at a site called a **TATA** box or **Hogness** box about −20 to −40 bp before the transcription start site. Once bound, TFIID helps organize other TFs required for initiation of RNA synthesis (see Figure 4-3b). The complex of transcription factors and RNA polymerase comprise the **preinitiation complex**, which yields basal levels of transcription. Induction to higher levels requires the presence of other activators binding to enhancer elements. Activator proteins like **Ctf**, **Sp1**, and **Oct-1** (see Figure 4-3c) cause the DNA to loop back upon itself so that the activator proteins interact with the preinitiation complex, signalling the RNA polymerase to begin synthesizing high levels of RNA.

RNA Processing

After transcription, RNA in eukaryotes undergoes significant processing. Transcripts that specify proteins are modified in the nucleus by the addition of 7-methylguanine **caps** at their 5' ends and **poly-A tails** approximately 100–250 nucleotides long at their 3' termini. The pre-mRNA is converted into mRNA by the excision of introns and splicing of exons. Most splicing is carried out by enzyme complexes, called **spliceosomes**, in the nucleus. Spliceosomes consist of 4 different **small nuclear ribonucleoprotein particles** (snRNPs) that work together to bring the ends of exons in a primary transcript near each other (see Figure 4-4). The snRNPs are constructed from six to ten proteins and one or two of the five **small nuclear RNAs** (snRNAs) designated **U1, U2, U4, U5,** and **U6.** The snRNPs are generally designated by the snRNAs they contain.

U1 snRNP binds to the 5′ exon-intron junction, U5 snRNP attaches near the 3′ intron-exon junction, whereas U4-U6 snRNP binds near U5, and U2 associates where a **lariat branch point** will form (see Figure 4-4b,c). The spliceosome, in particular U1 snRNP, cuts at the 3′ end of an exon (#5 as an example in Figure 4-4c). U2 snRNA catalyzes the formation of the lariat, whereas U5 catalyzes the cut at the 5′ end of exon 6 and the splicing of exon 5 to exon 6 (see Figure 4-4d).

In the simplest case, a spliceosome promotes the excision of an intron between two exons and the splicing together of the two exons. In more complicated cases, a spliceosome may promote **alternative splicing**, the splicing of a pre-mRNA into different combinations of targeted exons. The mRNA is subsequently transported to the cytoplasm where it is translated into proteins.

Solved Problems

Solved Problem 4.1 Why regulate operon induction?

If biosynthesis were unregulated, a cell would rapidly fill with mRNA molecules and enzymes whose activities may be unnecessary for cellular function at that time. This would be wasteful and counterproductive since the synthesis of these molecules requires a lot of energy which could be better spent in making repairs or for cell reproduction.

Solved Problem 4.2 How would mutations in the controlling sites affect an operon?

In the lactose operon, mutations in *lacO* have been discovered that alter repressor binding. These operator mutations, symbolized *lacO^c*, result in constitutive expression, even in the absence of inducer. Mutations in the *lacCRP* site which eliminate CRP binding reduce the expression of the operon. Mutations in the *lacP1* site have been discovered that reduce the expression by reducing RNA polymerase binding to *lacP1*.

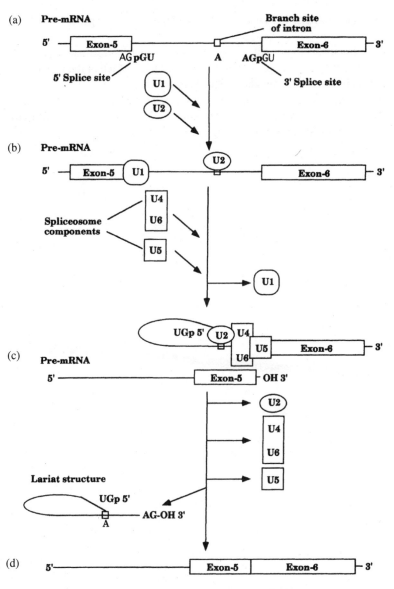

Figure 4-4 RNA splicing of pre-mRNA.

Solved Problem 4.3 Distinguish between introns and spacer DNA.

In more complex eukaryotes such as humans, it is estimated that over 95% of the genome consists of noncoding DNA. Much of the noncoding DNA is found *between* genes. This is referred to as spacer DNA. Introns are located *within* genes and are transcribed by RNA polymerase. They are later removed and the exons (coding regions) are spliced together to produce the final mRNA.

Chapter 5
TRANSLATION

The Genetic Code

A group of three adjacent nucleotides in DNA is transcribed into three complementary RNA nucleotides, which in turn are translated into a single amino acid within a polypeptide chain. Being a triplet code, $4^3 = 64$ different combinations exist, which is many more than are necessary to encode 20 different amino acids. Each coding triplet is referred to as a **codon**. Each mRNA codon in Table 5.1 is conventionally written with the 5' nucleotide at the left and the 3' nucleotide at the right, because protein synthesis begins at the 5' ends of mRNA molecules and proceeds toward their 3' ends.

The code is highly **degenerate** in that more than one codon can specify the same amino acid. Because of code degeneracy, many changes (**mutations**) can occur in a gene that will have no effect on the amino acid composition of the gene product. Such changes are referred to as **silent** mutations. The complementary base pairing between an mRNA codon and its **anticodon** in a tRNA is usually much less restrained at the third position than in the other two positions of the triplet. This phenomenon,

53

First Letter	Second Letter				Third Letter
	U	C	A	G	
U	UUU ⎱ Phe UUC ⎰ UUA ⎱ Leu UUG ⎰	UCU ⎱ UCC ⎱ Ser UCA ⎰ UCG ⎰	UAU ⎱ Tyr UAC ⎰ UAA ⎱ Nonsense UAG ⎰	UGU ⎱ Cys UGC ⎰ UGA Nonsense UGG Trp	U C A G
C	CUU ⎱ CUC ⎱ Leu CUA ⎰ CUG ⎰	CCU ⎱ CCC ⎱ Pro CCA ⎰ CCG ⎰	CAU ⎱ His CAC ⎰ CAA ⎱ Gln CAG ⎰	CGU ⎱ CGC ⎱ Arg CGA ⎰ CGG ⎰	U C A G
A	AUU ⎱ Ile AUC ⎰ AUA ⎰ AUG Met	ACU ⎱ ACC ⎱ Thr ACA ⎰ ACG ⎰	AAU ⎱ Asn AAC ⎰ AAA ⎱ Lys AAG ⎰	AGU ⎱ Ser AGC ⎰ AGA ⎱ Arg AGG ⎰	U C A G
G	GUU ⎱ GUC ⎱ Val GUA ⎰ GUG ⎰	GCU ⎱ GCC ⎱ Ala GCA ⎰ GCG ⎰	GAU ⎱ Asp GAC ⎰ GAA ⎱ Glu GAG ⎰	GGU ⎱ GGC ⎱ Gly GGA ⎰ GGG ⎰	U C A G

Table 5.1 Codons (displayed as mRNA triplets)

called **wobble**, allows the same tRNA to recognize more than one mRNA codon in many cases.

The codon 5'-AUG-3' near the end of an mRNA molecule is the usual **start (initiation) codon** that places methionine at the beginning (amino end) of all nascent eukaryotic polypeptide chains. Sixty-one codons are **sense codons** that specify amino acids. There are three codons that are not recognized by any tRNA: UAA, UAG, and UGA. These are termed **nonsense codons** or **stop codons**, because they provide part of the signal that protein synthesis should stop at that point. The completed polypeptide can be released from its cognate tRNA and from the ribosome.

Amazing!

The genetic code is **universal**! Essentially, the same codons encode the same amino acids in all organisms!

Translation in Prokaryotes

Ribosomes contain two sites for binding tRNA molecules: the **aminoacyl site** (A site), where each tRNA molecule first attaches, and the **peptidyl site** (P site), where a tRNA holds the growing polypeptide chain. The bacterial ribosome has two major subunits: a 30S subunit to which the mRNA and tRNAs become bound, and a 50S subunit to which tRNAs also bind. Each tRNA becomes **charged** by a **tRNA aminoacyl synthetase** that attaches a specific amino acid to each species of tRNA. Each tRNA molecule has a loop containing a triplet of ribonucleotides, called the **anticodon**, that can base pair with a complementary triplet codon in mRNA.

Translation in prokaryotes begins at a start codon. The bacterial mRNA translation initation codon (AUG) encodes N-formylmethionine, whereas internal AUG codons specify methionine. The 3'-UAC-5' anticodon of the tRNA pairs (in antiparallel fashion) with the complementary 5'-AUG-3' codon in the mRNA.

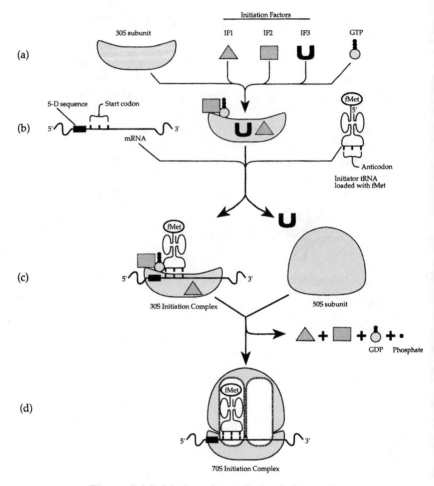

Figure 5-1 Initiation of translation in bacteria.

In the first step of translataion initiation, three protein **initiation factors** (IF1, IF2, IF3) and GTP bind to the 30S ribosomal subunit (see Figure 5-1a). The 30S subunit binds to a **Shine-Dalgarno** (S-D) recognition sequence on the mRNA due to base pairing interactions with the 16S rRNA component of the ribosome. Then, a tRNA loaded with the amino acid N-formylmethionine (tRNA^{fMet}) binds to the 30S-mRNA complex (see Figure 5-1b). IF3 is released and then GTP hydrolysis releases IF1,

IF2, GDP, and phosphate, allowing the 50S subunit to join the 30S-mRNA-tRNA^{fMet} complex to form a complete 70S **initiation complex** (see Figure 5-1c,d). The tRNA^{fMet} ends up in the P site of the ribosome. This completes the initiation phase.

The **elongation** phase proceeds with the help of a group of protein **elongation factors** when a second activated tRNA enters the A site (again by specific codon-anticodon base pairing). This places N-formylmethionine and the incoming amino acid next to one another so that a peptide bond can be formed between them by action of an enzymatic, ribosomal component called **peptidyl transferase**. The amino-acyl bond that held the N-formylmethionine to the 3′ end of its tRNA is broken simultaneously with the formation of the peptide bond. The mRNA translocates three nucleotides through the ribosome and a new codon is now positioned in the A site. This process is repeated until a stop codon is encountered at the A site (see Figure 5-2).

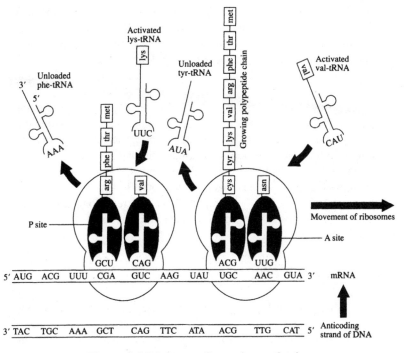

Figure 5-2 Diagram of protein synthesis.

At **termination**, the polypeptide is released from the tRNA with the help of **release factors**. The mRNA and the last tRNA are released from the ribosome as the ribosome dissociates into its 30S and 50S subunits.

Did You Know?

Many antibiotics block protein synthesis by binding to prokaryotic ribosomal subunits.

Translation in Eukaryotes

The process of translation in eukaryotes is essentially the same as that in bacteria, but differs in several important ways. Structurally, the ribosomal subunits of eukaryotes consist of 40S and 60S subunits that together form an 80S complex. Although most bacterial mRNAs specify multiple proteins, eukaryotic mRNAs code for a single nascent polypeptide chain. However, some newly synthesized polypetide chains may subsequently be enzymatically cleaved into two or more functional protein components.

Only three well-defined initiation factors are required for translation of *E. coli* mRNAs, but many more are needed in eukaryotes. During initiation, a special initiator tRNA (tRNA^{iMet}) brings an unformylated methionine into the first position on the ribosome. In eukaryotes the 40S ribosomal subunit is thought to attach at the capped 5′ terminus rather than at a Shine-Dalgarno sequence as in prokaryotes. It then slides along until it reaches the first AUG start codon. Three different elongation factors in eukaryotes replace those found in bacteria. However, a single release factors acts in eukaryotes.

Solved Problems

Solved Problem 5.1 Describe several characteristics of the genetic code.

The genetic code is triplet, meaning three nucleotides specify one amino acid. It is nonoverlapping, meaning three nucleotides are read, then the following three, and so on. The code is degenerate; more than one codon may encode the same amino acid. Lastly, it is universal; the same code is used in all living organisms, from bacteria to plants to animals.

Solved Problem 5.2 List the components of the prokaryotic translation initiation complex.

The prokaryotic translation initiation complex includes the 30S ribosomal subunit, fMet-tRNA^{fMet}, mRNA, 50S ribosomal subunit, protein initiation factors, and guanosine triphosphate (for energy).

Solved Problem 5.3 Are transcription and translation in eukaryotes coupled as they are in bacteria?

No. In bacteria, as soon as the 5' end of a bacterial mRNA is transcribed, ribosomes can attach and begin translation. However, in eukaryotes, the primary transcript must be processed into a functional mRNA molecule (see Chapter 4) and then must be transported from the nucleus to the cytoplasm before translation can begin.

Solved Problem 5.4 What amino acid sequence is encoded by the following mRNA?

5' GGAUGGAUUUUAAGUGAAG 3'

First you must find the start codon to set the reading frame. Then, set apart the codons in triplets. Use Table 5.1 to decipher the amino acid sequence.

5' GG <u>AUG</u> GAU UUU AAG UGA AG 3'

met-asp-phe-lys-STOP

Chapter 6
MUTATIONS

IN THIS CHAPTER:

✔ *Types of Mutations*
✔ *Mutagens*
✔ *Chromosomal Aberrations*
✔ *Solved Problems*

Types of Mutations

Mutations are heritable changes in the genetic material that give rise to alternative forms of any gene. These alternate forms are called **alleles**. There are two broad types of mutations, those that affect the gene and those that affect whole chromosomes (**chromosomal aberrations**). Gene mutations at the nucleotide level are called **point mutations**.

Any errors in the replication of a gene within the DNA molecule resulting in the insertion, deletion, or substitution of one or more bases will give rise to a mutation. Even though the cell has mechanisms to improve the fidelity of DNA replication, every once in a while a spontaneous mistake is made leading to a heritable change in the DNA sequence. In the laboratory, mutation rates can be greatly increased by exposing cells to chemicals or physical agents called **mutagens**.

Many mutations are due to the instability of the nucleotide bases in the DNA. Nucleotide bases may undergo structural changes called **tautomeric shifts** (see Figure 6-1), which result in the redistribution of electrons and protons so that the bases no longer pair normally. G may pair

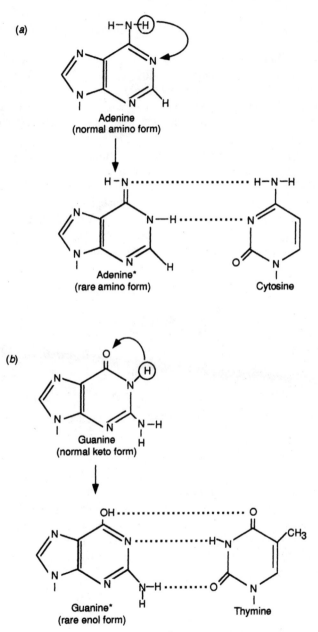

Figure 6-1 Tautomeric shifts and abnormal base pairing.

with T or A with C, resulting in heritable changes in the nucleotide sequence.

Transitions occur when the mispairing results in the replacement of one purine for another or one pyrimidine for another. **Transversions** result when a purine is replaced by a pyrimidine or vice versa. Because the structural changes leading to transitions are relatively small, they occur more frequently than transversions, which require more substantial modifications of a molecule.

Point mutations resulting from base substitutions in a gene that code for a polypeptide may result in **missense, nonsense,** or **silent** muations. Missense mutations result in the replacement of one sense codon for another, altering the amino acid encoded in that position. A nonsense mutation creates one of the three stop codons (UGA, UAA, UAG) and results in polypeptides that are shorter than the normal ones. Silent mutations are changes in the sequence of the codon that do not alter the encoded amino acid.

Note!

Sickle cell anemia results from a GAG→GUG transversion within a glutamine codon. The mutant hemoglobin protein has valine in its place, and these cells become crescent shaped under low oxygen tension.

Frameshift mutations are the result of nucleotide or base insertions or deletions within the coding region of a gene. The genetic code is translated by the protein synthesis apparatus by reading sequential groups of three bases that make up a codon, beginning from the start codon. If a single base is added or removed, all of the codons from that point on will be changed. A truncated protein may result if a codon is mutated to one of the three stop codons.

Mutagens

Chemical and physical mutagens can cause mutations by replacing one base with another in the DNA molecule, causing structural changes in a base so that it causes it to mispair, causing insertions or deletions, or damaging a base so much that it is unable to pair with any other normal base. **Base analogs** are sufficiently similar to the normal nitrogenous bases in DNA that they can be incorporated into a replicating DNA molecule by DNA polymerases. Once incorporated, however, base analogs have abnormal base-pairing properties so that they produce mutations during subsequent DNA replication cycles. For example, 5-bromouracil and 2-amino-purine are two common base analogs.

Alkylating agents cause mutations by chemically altering bases so that they pair up with a specific base other than the normally complementary base. **Intercalating agents** are planar molecules that can insert themselves between the stacked bases within the double helix. These agents alter the molecule of DNA in such a way that DNA polymerases may insert or skip one or more bases during replication, often resulting in frameshift mutations. For example, proflavin, acridine orange, and ethidium bromide are intercalators.

When ultraviolet light is absorbed by adjacent pyrimidines in one strand of a DNA molecule, a dimer forms. These dimers interfere with proper base pairing during DNA replication. The impact is so extensive that the normal replicative process is stopped until these dimers are repaired.

Chromosomal Aberrations

Two types of chromosomal aberrations can occur in cells: changes in chromosome structure and changes in chromosome number. Structural changes include deletions, duplications, inversions, and translocations (see Figure 6-2). **Deletions** are chromosomal changes in which one or more genes or chromosomal segments are lost. **Duplications** occur when one or more copies of a chromosomal segment are present on the same or different chromosomes. Deletions and duplications can occur in the same mutational event when two homologous DNA strands overlap, break at the same time at two different (nonhomologous) points, and then rejoin with the wrong strand. One of the strands will be missing one or

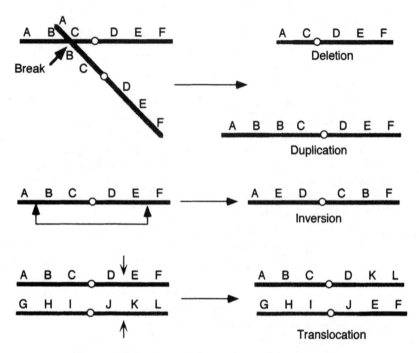

Figure 6-2 Structural aberrations of chromosomes.

more genes, and the reciprocal strand will have an extra copy of one or more genes. **Inversions** occur when a breakage in one of the chromosomes occurs and the segment rotates 180° before it rejoins. **Translocations** take place when nonhomologous chromosomes break and exchange segments.

In diploid (2n) organisms, there are two major types of chromosomal aberrations that are the result of changes in chromosome number. These are **polyploidy** and **aneuploidy**. Polyploidy results when cells acquire one or more sets of chromosomes beyond the "normal" number of sets. For example, **triploids** (3n) contain one extra set of chromosomes, and would therefore be sterile, since they cannot produce balanced gametes by meiosis.

Aneuploids are the result of changes in the individual number of homologous chromosomes in a set. This usually results from **nondisjunction** during meiosis (see Figure 6-3). The aneuploid condition that results in three copies of a given chromosome is known as **trisomy** (2n+1).

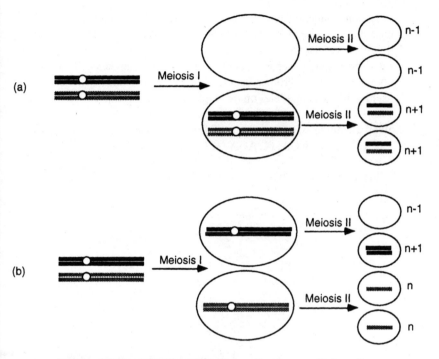

Figure 6-3 Aneuploid gamete formation by nondisjunction during (a) meiosis I and (b) meiosis II.

Solved Problems

Solved Problem 6.1 What would the result be if an adenine underwent a tautomeric shift as the DNA molecule was being replicated?

A cytosine would be incorporated into the new copy strand rather than a thymine. The other template strand, when replicated, will exhibit normal base pairing. When the DNA molecule with the mispaired region is replicated, a perfectly matched mutant DNA molecule will result and a G-C pairing will appear where an A-T was before.

Solved Problem 6.2 Distinguish between the resulting gametes from nondisjunction during meiosis I or II.

If nondisjunction occurs during meiosis I, homologous chromosomes will fail to separate. Half of the resultant gametes will be n−1 and half will be n+1. However, if nondisjunction occurs during meiosis II, sister chromatids fail to separate. Half of the resultant gametes will contain a normal number of chromosomes, and ¼ will be n−1 and ¼ will be n+1, with identical sets of alleles on the duplicated chromosome.

Solved Problem 6.3 Classify the following mutations: (1) A→T; (2) C→T; (3) AGA→UGA; (4) AGA→CGA; and (5) AGA→AAA.

(1) transversion; (2) transition; (3) nonsense; (4) silent; and (5) sense.

Chapter 7

BACTERIAL GENETICS AND BACTERIOPHAGES

IN THIS CHAPTER:

- ✔ *Bacteriophages*
- ✔ *Recombination*
- ✔ *Genetic Transfer*
- ✔ *Solved Problems*

Bacteriophages

A **bacteriophage** is a virus that infects bacteria. Like all viruses, phages are obligate intracellular parasites, devoid of protein synthesizing machinery and energy conversion systems. They contain nucleic acid enclosed in a protein coat, or **capsid**. Bacteriophages require a living bacterial host in which to carry out their reproductive cycle.

The **lytic** or **vegetative life cycle** culminates in lysis of the host cell and the release of numerous viral progeny. Bacterial viruses exhibiting only a lytic life cycle are known as **virulent** bacteriophages because they eventually cause the death and destruction of the host bacterium. For example, the T-even phages such as T2, T4, and T6 are virulent.

The life cycle of T4 is illustrated in Figure 7-1. The lytic cycle consists of five steps. The cycle begins by **attachment** of the bacteriophage

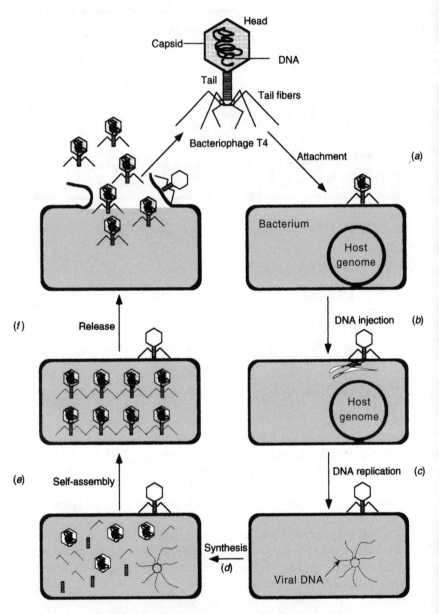

Figure 7-1 The lytic cycle of bacteriophage T4.

to molecules on the host's cell wall. Next, the virus introduces its genetic material into the cell (**penetration**). Once the bacteriophage genome enters the cytoplasm, it subverts the host's nucleic acid and protein synthesis apparatus and initiates the **synthesis** of viral proteins and DNA. As the viral proteins are synthesized, they self-assemble into viral components such as the head (containing the phage DNA), tail, and tail fibers. The **assembly** process results in the formation of numerous intact phage particles within the cell. After assembly, viral proteins cause lysis of the host cell, and all the viral progeny are **released** into the environment.

A **temperate** phage can cause a lytic infection, but can also exist within the host bacterium as a **prophage**, that is when the genetic material of the phage is inserted into the DNA of the host cell. In this state, the **lysogenized** bacteria can carry out a seemingly normal life cycle. When exposed to UV light or nutrient deprivation, the prophage is excised from the bacterial genome, leading to a lytic cycle.

Recombination

Genetic **recombination** in bacteria is a nonreciprocal process whereby segments of genetic material from two different sources are brought together into a single DNA molecule. Homologous recombination was discussed in Chapter 3. **Site-specific recombination** involves the recombination of two DNA molecules at specific locations variously called **insertion sequences** (IS), **long terminal repeats** (LTRs), and **attachment sites** (*att*). The integration of the bacteriophage λ into the *E. coli* chromosome is a common example of site-specific recombination involving *att* (see Figure 7-2). Both possess *att* sites, which are recognized by the λ integration and excision enzymes. Both chromosomes share a short region of homology indicated by "O." This region of homology is flanked by short DNA sequences that are unique to the organism. The flanking *E. coli* regions are indicated by B and B′, while those of the λ phage are indicated by P and P′. After integration of the phage DNA into that of *E. coli* by site specific recombination, the λ chromosome is flanked by the sequences BOP′ and POB′.

Genetic Transfer

Transformation is a mechanism of genetic transfer between bacteria in which the donor DNA exists cell-free in the recipient bacterium's imme-

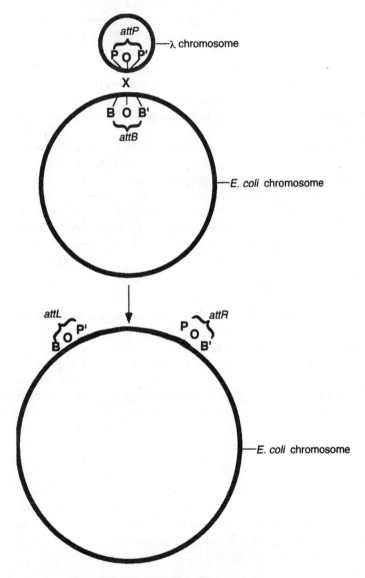

Figure 7-2 Integration of bacteriophage λ into the *E. coli* chromosome.

diate environment. DNA can be naturally released into the environment when cells die and subsequently lyse. Experimentally, DNA containing genes of interest, usually within a plasmid, can be introduced into the environment in order to transform bacterial cells. The ability of a recipient bacterium to take up free DNA and become transformed is known as **competence**. Some strains of bacteria are naturally competent. In others, competence is a brief physiological state during the exponential growth phase; in these bacteria, Ca^{2+} ions enhance the level of competence.

Transduction is a mechanism of DNA uptake by bacteria in which the donor DNA, consisting of fragments of the bacterial chromosome, is introduced into a bacterial cell via a phage vector. In **generalized transduction**, virtually any bacterial gene can be transferred by a lytic bacteriophage. During packaging of the viral DNA into the capsids, some of the host's DNA may be packed into the virus along with an incomplete viral genome. This virus will be able to initiate infection, and therefore introduce the original host's DNA into a new bacterial cell, but will not be able to replicate itself or lyse the new host cell.

Specialized transduction is a process whereby a lysogenic bacteriophage serves to transfer a specific gene at a high frequency. When lysogenic bacteriophages infect host cells, their DNA is incorporated into the host's genome by site-specific recombination, which always occurs at a specified location and adjacent to certain genes. Through the process of **induction**, the prophage genome becomes excised from the host and undergoes a lytic cycle. Occasionally, the phage excision from the host's genome is defective and results in the release of a viral genome that contains part of the host's genome, in particular, those genes that are adjacent to the phage's site of insertion.

Conjugation is a process during which genetic information is transferred unidirectionally from a donor bacterium to a recipient through a cytoplasmic channel between the two cells. The required cell-to-cell contact between the donor and recipient can be achieved through **sex pili**, through **agglutinins** (substances that promote cell clumping), or via **pheromones** (chemicals that alter the behavior of other members of the same species).

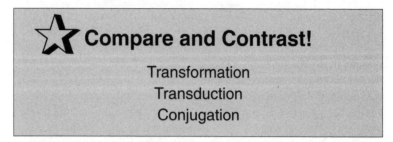

Compare and Contrast!

Transformation
Transduction
Conjugation

There are pieces of DNA measuring 700–20,000 bp in length that "jump" from one region of the genome to another. In both prokaryotes and eukaryotes, the jumping can occur from one location on the chromosome to another or, in the case of bacteria, from the chromosome to a plasmid, or vice versa. These "jumping genes" are called **transposable elements** or **transposons**. They are of importance since they may turn genes on or off when they move from one location to the control region of another gene, and they may cause mutations if they insert themselves within a gene. During the process, **transposase** causes breaks in both the transposon itself and at sites adjacent to the target sequence in the bacterial genome. The transposable element inserts itself by site-specific recombination into the genome and enzymes repair the gaps.

Solved Problems

Solved Problem 7.1 What mechanism of genetic exchange could not occur in a culture medium containing DNA?

Transformation could not occur, since it depends on the uptake of free DNA in the immediate environment.

Solved Problem 7.2 Can virulent phages assume a prophage state?

No. Only lysogenic viruses can assume a prophage state, that is, a state where the viral genome is inserted into the host chromosome. By definition, virulent phages undergo the lytic cycle.

Chapter 8
RECOMBINANT DNA TECHNOLOGY

IN THIS CHAPTER:

✔ *Cloning*
✔ *Restriction Endonucleases*
✔ *Vectors*
✔ *Host Cells*
✔ *Solved Problems*

Cloning

Discoveries in molecular biology have allowed scientists to duplicate natural genetic transfer phenom-
ena in the laboratory and to develop methods to introduce almost any type of genetic information into
an organism. **Genetic engineering** is the creation of
new DNA, usually by linking DNA from different or-

ganisms together by artificial means using enzymes known as **restriction
enzymes**. **Cloning** is the production of many copies of the newly engi-
neered DNA. The amplification of a specific cloned gene or genes, cou-
pled with a marked increase in production of their protein products, makes it relatively easy to extract and purify these proteins in the laboratory.

A typical cloning procedure is illustrated in Figure 8-1. A suitable
plasmid (vector) is selected in which to insert a desired gene (donor

73

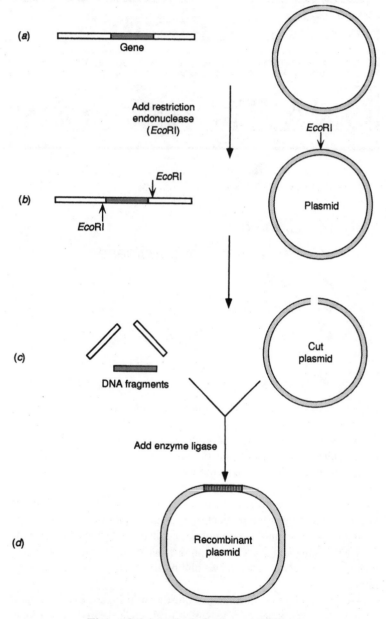

Figure 8-1 A typical cloning procedure.

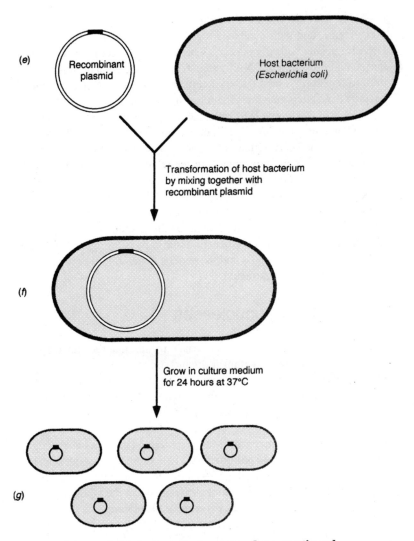

Figure 8-1 A typical cloning procedure, continued.

DNA). Both donor DNA and vector are digested with the same restriction enzyme, and then incubated together with ligase to join the donor DNA fragments with the plasmid. The result is a recombinant plasmid that contains the desired DNA fragment. The recombinant plasmid is then used to transform a host bacterial cell, creating a new genetic strain of bacteria that stably maintains the recombinant plasmid.

The goal of cloning is to isolate a desired gene or segment of DNA from an organism and introduce it into a suitable host cell to obtain large quantities of the DNA. Often, this donor DNA is used for the large-scale production of important proteins, but the DNA may also be used in the detection of infectious agents or abnormal cells. Normally, the donor DNA is a small portion of the genome of a cell, and it is present as one or two copies in each cell. Therefore, before donor DNA can be extracted, a sufficient number of cells containing the desired DNA must be obtained, either from a small segment of tissue or by culturing the cells. The cells must then be disrupted and the genetic material (either in chromosomes or in plasmids) extracted.

Restriction Endonucleases

Restriction endonucleases are bacterial enzymes that recognize specific nucleotide sequences within a double-stranded DNA molecule and cleave at those locations. These enzymes cut DNA into fragments of various lengths, depending on the number of times the enzyme's recognition site is repeated within the molecule. Most restriction endonucleases used in cloning recognize base pair sequences four to eight nucleotides long and cut within these sequences. Many restriction enzymes have recognition sequences known as **palindromes**. Palindromes are sequences that are identical when read in the $5' \rightarrow 3'$ direction on both strands of the DNA molecule. Restriction endonucleases may cut the DNA to produce fragments with **cohesive** ("sticky") **ends** or **blunt ends** (see Figure 8-2).

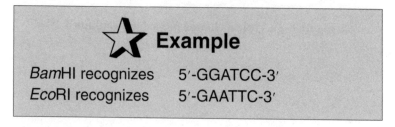

⭐ **Example**

*Bam*HI recognizes	5'-GGATCC-3'
*Eco*RI recognizes	5'-GAATTC-3'

(a) Cohesive ends are formed when *Bam*HI **cleaves the DNA**

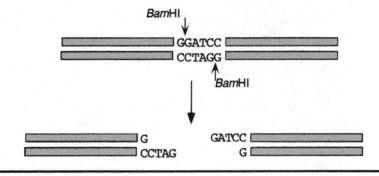

(b) Blunt ends are formed when *Hae*III **cleaves the DNA**

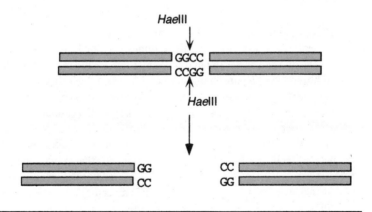

**Figure 8-2 Restriction enzymes may form (a) cohesive
or (b) blunt ends.**

The resulting fragments of a restriction enzyme digestion can be visualized by a procedure known as **electrophoresis**. Electrophoresis involves the movement of charged molecules or ions through a semisolid support medium under the influence of an electrical field. Agarose gels are common media for the electrophoresis of DNA. The agarose gel, cast as a thin slab in a mold with sample wells at one end, is submerged in a buffer solution with the sample well side toward the negative pole (**cath-**

ode). The samples are dispensed into the wells and a current from the power supply is applied to the system. Since nucleic acids have a negative charge at pH=8.0, they migrate within the gel matrix from the negative to the positive pole (**anode**) at a rate dependent upon their size, shape, and total charge. DNA molecules are invisible to the naked eye, but can be seen in gels by staining them with a solution of a dye called **ethidium bromide**, which intercalates between the stacked bases of the DNA molecule and fluoresces.

The graphical representation of recognition sites for two or more restriction endonucleases is known as a **restriction map** for that molecule. Knowing the restriction maps for commonly used plasmids and bacteriophage DNA genomes allows scientists to plan cloning stategies for isolating and moving around pieces of DNA that contain genes of interest.

Vectors

After a desired segment of DNA is cut away from the donor's genome with restriction endonucleases, it is **ligated** into a **vector** DNA molecule, usually a plasmid or a bacteriophage genome. **Ligases** are enzymes that catalyze the formation of a phosphodiester bond between the 3'-hydroxyl group of a segment of donor DNA and the 5'-phosphate group of the vector DNA. A vector is a DNA molecule into which foreign DNA moleules are ligated and inserted into cells so that the recombinant DNA can be replicated. Plasmid vectors must also contain a marker, such as an antibiotic resistance gene, to facilitate the selection of bacterial cells that contain the plasmid. Vectors can be introduced into host cells by transformation or transduction (see Chapter 7). An **expression vector** is a vector that carries a gene that can be efficiently transcribed and translated by the host cell.

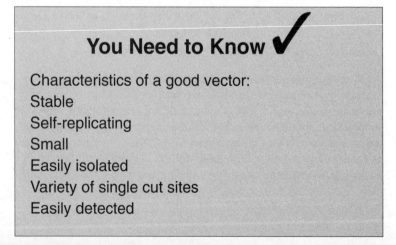

You Need to Know ✔

Characteristics of a good vector:
Stable
Self-replicating
Small
Easily isolated
Variety of single cut sites
Easily detected

Host Cells

A number of bacterial and yeast strains have been developed for recombinant DNA experiments. In order for a given plasmid to be replicated by a host cell, the cell must recognize its origin of replication site (*oriC*). Recombinant plasmid vectors are normally introduced into competent cells by transformation and then selected using appropriate cell culture media. For example, if the vector contains an *amp^R* gene that encodes resistance to ampicillin, the culture media would include that antibiotic to ensure that only transformed cells will grow.

Another method for introducing recombinant DNA molecules into host cells is **electroporation**. In this method, a suspension of exponentially growing host cells is mixed with a solution of recombinant DNA molecules and exposed to a high electric field for a few milliseconds. The high voltage alters the structure of the membrane so that pores are temporarily formed, allowing plasmid DNA to enter the cell. This method is fast and efficient.

If bacteria are used as the host to clone eukaryotic genes, certain steps must be taken to make it possible for the bacteria to make sensible mRNA and functional proteins, since bacteria do not possess mechanisms for processing eukaryotic pre-mRNA molecules. To do this, it is necessary to isolate already processed mRNA from the donor eukaryotic cells and convert the single-stranded RNA to double-stranded DNA. **Reverse transcriptase** from retroviruses (see Chapter 10) uses RNA templates for synthesizing DNA. The resulting DNA molecules, known as cDNA (for complementary DNA), can then be used for cloning in bacteria since they posses only intron-free protein-coding genetic information.

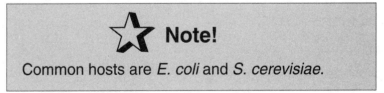

⭐ Note!
Common hosts are *E. coli* and *S. cerevisiae*.

Solved Problems

Solved Problem 8.1 How many fragments would be generated by a restriction endonuclease in a plasmid that has two recognition sequences?

Because plasmids are circular molecules, two fragments would be generated. However, if the DNA were linear and contained two recognition sequences, three fragments would be generated.

Solved Problem 8.2 Suppose the restriction endonuclease *Hind*III cuts a 6.0 kb linear piece of DNA into two fragments; an 800 bp fragment and a 5200 bp fragment. *Nar*I cuts the DNA also into two fragments; fragments 1200 and 4800 bp long. Relative to the *Hind*III cut site, there are two possible ways in which *Nar*I could have cut the DNA. How can you determine the correct cleavage site for *Nar*I with respect to *Hind*III?

To determine this, one must subject the DNA to a double digest in which both the enzymes are allowed to cut the DNA at the same time. When the double digest is allowed to take place, if the three fragments that appear upon electrophoresis of the restricted DNA are 400, 800, and 4800 bp, then the only possible way in which the data can be interpreted is with the 1200 bp *Nar*I fragment containing the *Hind*III recognition site 800 bp from the end of the linear piece of DNA. If the 4800 bp *Nar*I fragment contained the cut site, you would visualize fragments of sizes 800, 1200, and 4000 bp after electrophoresing the doubly digested DNA.

Chapter 9
NUCLEIC ACID MANIPULATIONS

IN THIS CHAPTER:

- ✔ *Nucleic Acid Hybridization*
- ✔ *The Polymerase Chain Reaction*
- ✔ *Nucleic Acid Sequencing*
- ✔ *Solved Problems*

Nucleic Acid Hybridization

From developments in the area of genetic engineering and molecular biology, a powerful tool known as **DNA hybridization** has emerged. This technique is used to detect the presence of DNA from pathogens in clinical specimens and to locate specific genes in cells. DNA hybridization takes advantage of the ability of nucleic acids to form stable, double-stranded molecules when two single strands with complementary bases are brought together under favorable conditions.

In DNA hybridization assays, DNA from a virus or cell is denatured with alkali to separate the strands. The single strands of DNA are then attached to a solid support such as a nitrocellulose or nylon membrane so that the strands do not reanneal (see Figure 9-1). The DNA is attached to the membrane by its sugar-phosphate backbone with the nitrogenous bases projecting outward. To characterize or identify the target DNA, a

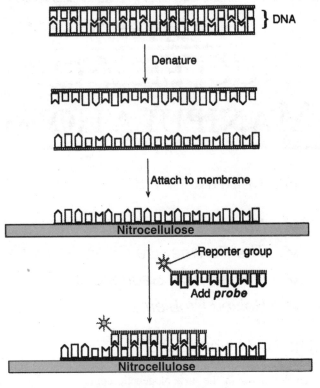

Figure 9-1 DNA hybridization.

single-stranded DNA or RNA molecule of known origin, called a **probe**, is added to the membrane in a buffered solution. This allows the formation of hydrogen bonds between complementary bases. The probe, so called because it is used to seek or probe for DNA sequences, is labeled with a **reporter** group, which may be a radioactive atom or an enzyme whose presence can be easily detected.

The probe is allowed to react with the target DNA; then any unreacted probe is removed by washing in buffered solutions. After the washes, all that remains on the nitrocellulose is the target DNA and any probe molecules that have attached to complementary sequences in the target DNA, forming stable hybrids.

Hybridization of target and probe DNAs is detected by assaying for the probe's reporter group. If the reporter group is detected, hybridization has taken place. If no reporter group is detected, it can be assumed that the target molecule does not have sequences that are complementary to those of the probe, and hence, the gene or DNA segment sought is not present in the sample.

Remember

Four components of DNA hybridization:

1. Target DNA
2. Probe
3. Detection system
4. Format

Three common formats are used in solid-phase hybridization assays; dot blot, Southern blotting, and *in situ* hybridization. In the **dot blot** assay, a specified volume of sample or specimen is spotted onto a small area of nitrocellulose membrane, which is then carried through the procedure described above. **Southern hybridization** assays (Figure 9-2) involve restriction enzyme digestion and agarose gel electrophoresis of the target DNA prior to the hybridization assay. The different bands on the agarose gel are transferred by capillary action onto a nitrocellulose or nylon membrane in a blotting apparatus. During the transfer, each of the DNA bands is transferred onto the membrane in the same relative position that it had in the gel. After the transfer, the target DNA is probed and detected, as in the dot blot assay. *In situ* **hybridization** assays involve the probing of intact cells or tissue sections affixed to a microscope slide. This type of solid-phase assay has the advantage that one cannot only detect the presence of target DNA in intact cells but also determine the location of such target DNA within a tissue. An important application of *in situ* hybridization is for the detection of viruses and certain types of bacteria within infected cells.

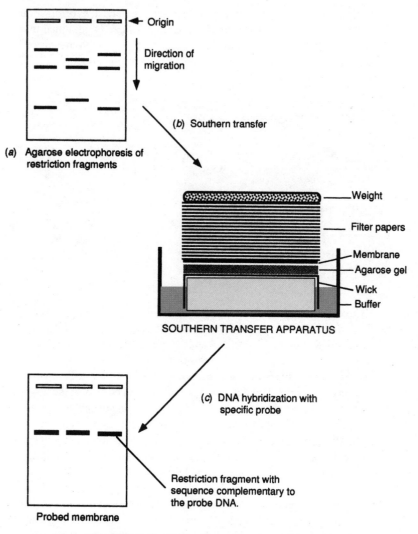

(a) Agarose electrophoresis of restriction fragments

(b) Southern transfer

SOUTHERN TRANSFER APPARATUS

(c) DNA hybridization with specific probe

Restriction fragment with sequence complementary to the probe DNA.

Probed membrane

Figure 9-2 Southern hybridization analysis.

The Polymerase Chain Reaction

The replication of genetic material is carried out by enzymes called DNA polymerases. These enzymes initiate the synthesis of DNA starting from a primer bound to a template (Chapter 3). The primers are generally 9 to 25 bases in length and establish the site where DNA replication begins. With the **polymerase chain reaction** (PCR), any particular stretch of genetic material can be pinpointed and replicated numerous times simply by selecting a pair of primers that flank the desired stretch of DNA. The PCR is predicated on the annealing of two oligonucleotides (primers) of known composition to a target sequence of interest and the extension of the oligonucleotides with a DNA polymerase. Each reaction is repeated subsequent to a denaturation step, thus allowing for exponential amplification.

The PCR (see Figure 9-3) involves three temperature incubations or steps that are repeated 20-50 times. One repetition of three steps is called a **cycle**. In the first step, called **denaturation**, the two strands of the target DNA molecule are separated (denatured) by heating the DNA to 94°C to break the hydrogen bonds between bases, yielding two separate strands. In the second step, called **annealing**, two primers hybridize to complementary sequences in the single strands. The primers are short (20–30 bases in length), synthetic stretches of single-stranded DNA. They are selected so that one primer is complementary to one end of the gene of interest on one strand, while the second primer is complementary to the opposite end on the other strand. The primers form hydrogen bonds with (anneal to) their complementary sequences, forming stable, double-stranded molecules. Annealing temperatures range between 37 and 60°C. During the third step, **extension**, or **elongation**, the primers are extended by a thermostable DNA polymerase at 72°C.

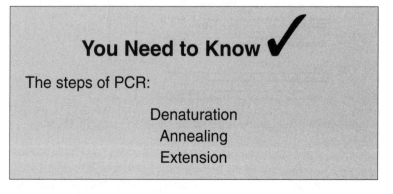

You Need to Know ✔

The steps of PCR:

Denaturation
Annealing
Extension

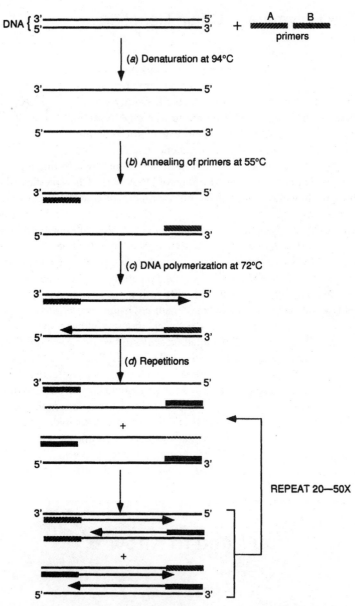

Figure 9-3 Schematic representation of a typical PCR assay.

To study the effects of mutations on gene expression, researchers have developed a technique known as **site-directed mutagenesis**, which introduces point mutations at specific sites. One of the most commonly used strategies takes advantage of primer-directed amplification of DNA to introduce mutations. One of the primers is designed with a sequence complementary to the region in the target DNA, but with the desired substitution, insertion, or deletion. The mutagenic sequence within the primer must be either at the 5' end of the primer or internal to the primer, but never at the 3' end of the mutagenic primer. The 3' end of the mutagenic primer (at least 6–10 bp long) must be totally complementary to the target DNA to permit full annealing of the primer to its target and allow the polymerase to extend the primer. The PCR is carried out initally (first 5–10 cycles) under low stringency conditions, to allow the mismatch to occur. Once a few mutagenized templates are produced during the PCR, these will serve as targets and will be fully complementary to the primer. The end products will contain the mutation at the desired site.

Nucleic Acid Sequencing

Nucleic acid sequencing reveals the genetic code of a DNA molecule. It may be carried out using one of two methods, each of which results in the production of DNA fragments of various lengths, differing from each other by a single base and from which one can infer the nucleic acid sequence of the molecule. This is accomplished using **denaturing polyacrylamide gels**. Whereas agarose gels can separate DNA molecules differing in length by 30–50 bases, polyacrylamide gels can discriminate among DNA molecules differing in length by a single base. Denaturing gels cause the DNA molecule to become single stranded and remain that way throughout the entire process of electrophoresis. Denaturing gels contain **urea** and are run at elevated temperatures, both of which promote the separation of the two strands of the DNA molecule.

Again, the DNA must be labeled in order to be visualized. The most common form of labeling is with radioactive isotopes, in particular, ^{32}P, ^{33}P, or ^{35}S. After electrophoresis, the gel is dried and placed next to a sheet of x-ray film in a dark place. During this time the radioactive particles emitted from the isotope in each DNA molecule "expose" the film, and after development, a dark band is seen on the film at the position where the DNA band was located in the gel. This picture, called an **auto-**

radiograph, is a mirror image of the position of the DNA bands in the gel.

There are two methods that can be used to sequence DNA molecules. The **Maxam-Gilbert** method is based on cleavage of DNA at specific sites by chemicals rather than enzymes. However, this method is seldom used anymore; the **Sanger** method is preferred.

In the Sanger method, the enzymatic synthesis of DNA takes place by the sequential formation of a phosphodiester bond between the free 5′ phosphate group of an incoming nucleotide and the 3′ OH group of the growing chain. This process takes place throughout the length of the DNA molecule. **Dideoxynucleotides** lack a 3′ OH group, and have a 3′ H group instead. In the presence of a dideoxynucleotide, the synthesis of DNA stalls because the diphosphate bond cannot be formed. The chain growth terminates at that point, and the last base on the 3′ end of the chain is a dideoxy terminator. This modification of Sanger's method of DNA sequencing is known as **dideoxy termination sequencing**.

Remember!

Maxam-Gilbert → chemical
Sanger → enzymatic

In the Sanger sequencing technique, four different reaction mixtures are used to sequence a DNA fragment. Each reaction mixture contains the template DNA molecule to be sequenced, radioactively labeled primers, all four deoxynucleotides, DNA polymerase, and a different dideoxy terminator (ddATP, ddCTP, ddGTP, or ddTTP). When one of these terminators is incorporated in the newly synthesized DNA strand, it will stop further synthesis of that strand; the result is that all the strands of various lengths in the reaction mixture end in the same base. The radioactive products are separated by electrophoresis and visualized by autoradiography. Reading from the bottom of the gel (shortest fragments terminated closest to the 5′ end) upward reveals the base sequence complementary to that of the template strand.

Solved Problems

Solved Problem 9.1 What components would one need to add to a PCR reaction?

In a typical PCR assay, an excess of primers, nucleotide triphophates, target DNA, a thermostable DNA polymerase, and a buffer containing appropriate salts and ions are added.

Solved Problem 9.2 What is the sequence of the template strand used to generate the following autoradiograph by the Sanger method of DNA sequencing?

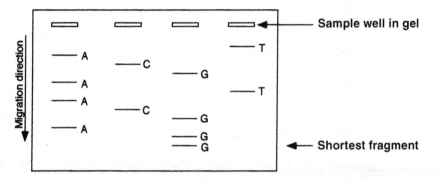

Reading from the bottom upward, the sequence represented on the gel is 5'-GGAGCATAGCAT-3'. Thus, the complementary sequence on the template would have been 5'-ATGCTATGCTCC-3'.

Solved Problem 9.3 Which DNA hybridization assay format is most useful when one wishes to: (1) detect the presence of a pathogen's DNA in an aqueous clinical sample; (2) detect the presence and location of a gene segment in a restriction digest of genomic DNA; and (3) detect the presence and localization of a pathogen's DNA within a cell?

The most useful assay formats would be (1) dot blot; (2) Southern blot; and (3) *in situ* hybridization.

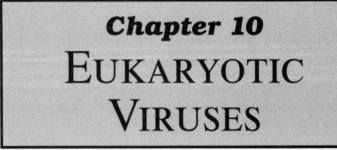

IN THIS CHAPTER:

- ✔ *Viral Structure*
- ✔ *Animal Viruses*
- ✔ *Oncogenic Viruses*
- ✔ *Plant Viruses*
- ✔ *Solved Problems*

Viral Structure

All viruses are noncellular infectious agents that proliferate only within cells. Although eukaryotic cells and their viruses carry out many of the same processes of bacteriophages (Chapter 7), the details of these processes differ, especially those carried out in specialized organelles. Some important processes are found almost exclusively in eukaryotes and their viruses; among these are RNA processing (exon splicing) and protein modifications (proteolytic cleavage, glycosylation, and phosphorylation).

Viruses that infect eukaryotic cells consist of a nucleic acid, either DNA or RNA, covered by a protein coat, a **capsid**. A single protein subunit of the capsid is referred to as a **capsomere**. The capsids of most eukaryotic viruses consist of a number of different proteins. The complex of nucleic acid and capsid is designated the **nucleocapsid**. Many animal viruses are surrounded by a membrane (lipid bilayer) derived from the host cell in which they proliferate. These viruses are said to be **enveloped**. The complete, intact virus particle is referred to as a **virion**.

⭐ Note!
Viruses are not considered "living" organisms.

The primary characteristics used to differentiate eukaryotic viruses are associated with their nucleic acid. First, viruses may be separated based on whether they are **DNA viruses** or **RNA viruses**. The nucleic acid may be single-stranded (ss) or double-stranded (ds), depending on the species. If the ssRNA is able to function as mRNA it is referred to as **plus strand RNA** (+RNA); if it is the equivalent to antisense RNA it is known as **minus strand RNA** (−RNA). Some of the genomes in plant and animal viruses are fragmented into segments. Virion shape is also used to differentiate among the viruses since they have a number of distinctive forms: cylindrical or helical, spherical, icosahedral, bullet-shaped, or even more complex shapes. The presence or absence of an envelope and the virion size are also helpful in distinguishing viruses.

Animal Viruses

An overview of eukaryotic viral **infection** is illustrated in Figure 10-1. For most animal viruses, the first step toward proliferation is **attachment** (or **adsorption**) to the surface of a host cell. This is mediated by specific proteins associated with the capsid or envelope, referred to as **viral binding sites** or **viral attachment proteins**. The attachment proteins interact with specific proteins or polysaccharides on the surface of the host cell. The proteins on the host cell are called **receptors**. The specificity of viral attachment proteins and of the host receptors determine to which host cells viruses can adsorb and subsequently infect.

Once viruses have attached, they **penetrate** across the host cell's plasma membrane by either **endocytosis** or **membrane fusion**. In endocytosis, viruses enter the cytoplasm within **endosomes** derived from the invagination of the plasma membrane. A lowering of pH in endosomes is

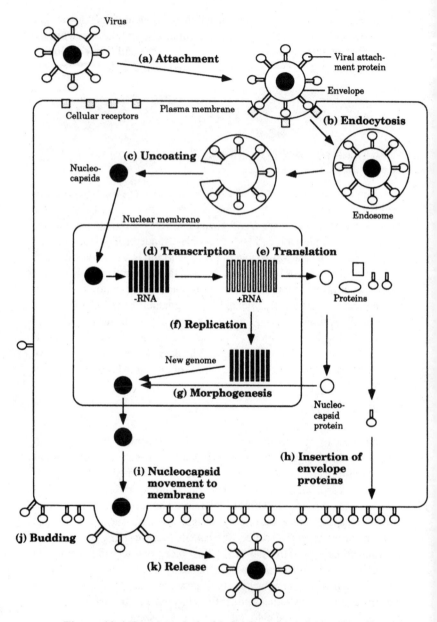

Figure 10-1 Enveloped viral infection of a eukaryotic cell.

frequently the trigger that leads to capsid disintegration and nucleic acid release, a process known as **uncoating**. The genomes of most DNA viruses find their way to the nucleus by an unknown mechanism. In the nucleus, the genome is replicated. The RNA viruses generally replicate in the cytoplasm, but there are exceptions including the unique case of retroviruses, discussed below.

Once the viral nucleic acid reaches the appropriate compartment, it is transcribed and the transcripts are translated. This usually happens before nucleic acid replication is initiated, since some of the proteins synthesized early in infection play a role in viral nucleic acid replication. Following genome replication, the nucleic acid becomes packaged, the virions are assembled, and then **release** occurs. In nonenveloped viruses, accumulation of virus particles in the cytoplasm causes the host cell to rupture. Assembly of enveloped viruses is closely tied to their release. Nucleocapsids binding to viral proteins in the plasma membrane trigger **budding**, resulting in enveloped virions.

The replication of the nucleic acid is extremely diverse, however, some general features can be outlined. Single-stranded +RNA viruses, such as picornviruses and togaviruses, maybe directly translated. A viral-encoded RNA-dependent RNA polymerase is produced early on and catalyzes the transcription of intermediate complementary −RNA strands. The minus strands then serve as templates for the synthesis of genomic +RNAs.

Retroviruses are an unusual group of +RNA viruses in that they synthesize new +RNA using a DNA template. The plus genome is converted in a step-wise manner to a dsDNA molecule by **reverse transcriptase**, which is carried by the virus. A second enzyme, **ribonuclease H**, digests the RNA in the intermediate RNA-DNA hybrid. Reverse transcriptase subsequently synthesizes a DNA strand complementary to the first and the resulting dsDNA integrates into the host genome.

Did You Know?

HIV, which causes AIDS, is a retrovirus.

Single-stranded −RNA viruses are those with nucleic acids (genomes) that cannot be translated. The −RNA strand serves as a template for the synthesis of +RNA that functions as mRNA. Viral proteins translated from the mRNA promote the synthesis of full-length +RNA strands that function as templates for the synthesis of full-length –RNA genome strands. Usually the virus brings into the cell an RNA-dependent RNA polymerase for making viral mRNA.

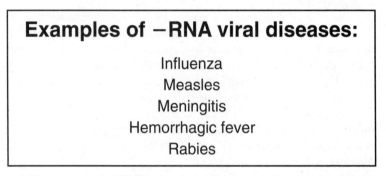

Examples of − RNA viral diseases:

Influenza
Measles
Meningitis
Hemorrhagic fever
Rabies

Double-stranded RNA viruses, such as reoviruses, are segmented and carry a dsRNA-dependent RNA polymerase which they use to transcribe their genomes. The RNA segments and the mRNA molecules specified by them only encode for one protein each. These viruses replicate their dsRNA genomes by producing many copies of plus strand RNA that are not translated, but serve as templates for the synthesis of complementary minus strands.

Single-stranded DNA viruses are unusual in that that may have either **sense** or **antisense** DNA. Sense DNA can serve as a template for mRNA synthesis, whereas antisense DNA is unable to function as such a template. Sense DNA can be transcribed immediately, but antisense DNA must serve as the template for the synthesis of the sense strand.

The dsDNA viruses can be divided into two categories: those that replicate their DNA in the host's nucleus and those that replicate entirely in the host's cytoplasm (the poxviruses). Although there are differences in the processes of replication, generally, the viral DNA has to be transcribed and viral proteins synthesized for DNA replication to occur.

> # Examples of dsDNA viruses:
>
> Adenoviruses
> Herpesviruses
> Papillomaviruses
> Poxviruses

Oncogenic Viruses

Oncogenes are mutated or inappropriately expressed cellular genes that specify proteins in signal transduction pathways. **Oncoproteins** are the oncogenic gene products that function in signal transduction pathways to **transform** cells, allowing them to proliferate in an uncontrolled manner, generally resulting in cancerous growths (e.g., malignant tumors in solid tissues).

Viruses such as retroviruses carry oncogenes derived from normal cellular genes (**proto-oncogenes**) and are called **oncoviruses**. They also contain viral genes that promote cellular proliferation, thus producing more cells in which mutations in proto-oncogenes might occur. These viruses transform cells when they integrate to become **proviruses** and disrupt tumor suppressor genes or cause inappropriate expression of normal proto-oncogenes. Transformed cells often proliferate out of control, become immortal, change shape, have new antigenic properties, and lose **contact inhibition**. Normal cells usually stop proliferating when sufficient contacts have been made with other cells. The loss of contact inhibition allows cells to wander off into other tissues and organs (**metastasis**) and spread a cancerous growth. In some cases, the oncogenic protein is an overproduced but normal protein, but mostly oncoproteins differ from the normal protein in their amino acid sequence. If the oncoprotein is part of a signal transduction pathway, then the abnormal component stimulates the cell to replicate itself inappropriately. This disruptive transformation may, after several other mutations, yield cancerous cells.

Oncoproteins fit into one of eight categories: (1) peptide growth factors, (2) growth factor receptors in the plasma membrane or cytoplasm, (3) GTP-regulated proteins called G proteins (see Chapter 11), (4) membrane receptors with tyrosine kinase or with threonine-serine kinase activities, (5) cytoplasmic protein kinases with tyrosine kinase activities or with serine-threonine activities, (6) DNA-binding proteins that function as transcriptional activators or that promote DNA replication, (7) cyclins that promote the activity of protein kinases, and (8) proteins that inhibit tumor suppressor proteins.

Almost all of the oncoproteins function in various signal transduction pathways that begin with a signal (peptide or steroid hormone) and end with the activation of transcription and/or the initiation of DNA replication. The oncoproteins override the normal regulation of cells and continuously send signals that activate gene expression and progression through the cell cycle. This increases the chances that mutations will occur in the proto-oncogenes and in **tumor suppressor genes** (normal cellular genes whose products dampen or inhibit cell replication). The more proto-oncogenes converted into oncogenes, the more unregulated a cell becomes. Similarly, the more tumor suppressor genes that are damaged by mutations, the more signal transduction pathways or cell cycle regulation mechanisms that do not function properly.

Examples of oncogenes:

Ras	*Myb*
Src	*Abl*
Fos	*Jun*

Plant Viruses

While there are exceptions, the vast majority of plant viruses have a single-stranded linear, +RNA genome with a capsid having helical or icosahedral symmetry. They have small genomes, encoding for only three to four proteins: (1) a helicase; (2) an RNA replicase; (3) a cell-to-cell movement protein; and (4) a capsomere. The helicase is thought to be important in the unwinding and separation of the plus and minus RNA strands. The replicase, an RNA-dependent RNA polymerase, is encoded in those viruses that are able to use the host's enzymes. The cell-to-cell movement protein facilitates the spread of the viral RNA through plant tissue. The capsomere is the protein subunit of the capsid.

Many plant viruses depend on insect vectors to infect plant cells. Tobacco mosaic virus is only dependent on mechanical damage to cell walls, which allows the virus to bind to the plasma membrane of host cells.

Solved Problems

Solved Problem 10.1 Describe the capsid of an icosahedral virus.

Icosahedral virsues have a capsid with 20 triangular faces, 12 vertices, and 30 edges. The simplest icosahedral viruses have faces composed of three capsomeres. Such a capsid has a total of 60 capsomeres. Many viral capsids with more than 20 faces but having icosahedral symmetry are also referred to as icosahedrons.

Solved Problem 10.2 True or false? (1) Single-stranded +RNA viruses usually engage in translation before they transcribe. (2) Retroviruses are single-stranded +RNA viruses, and they carry a DNA polymerase that can only use DNA templates. (3) Generally, viral adsorption is mediated by lipid attachment sites in the virus and lipid receptors on the host cell.

(1) True. (2) False, the DNA polymerase reverse transcriptase can use either RNA or DNA as a template. (3) False, adsorption is mediated by protein attachment sites/protein receptors.

Chapter 11
CELL
COMMUNICATION

IN THIS CHAPTER:

✔ *Introduction*
✔ *G Proteins*
✔ *Kinases and Phosphatases*
✔ *The Cell Cycle*
✔ *Solved Problems*

Introduction

Cells are continually receiving information from their surroundings, and they must be able to respond appropriately. Most extracellular chemical signals fall into one of three categories: (1) proteins and peptides; (2) peptide neurotransmitters; (3) and steroids and other membrane soluble molecules. Physical signals such as electromagnetic radiation (light) and heat are also important. Growth, proliferation, differentiation, movement, and programmed cellular death all depend upon signals altering a cell's physiology, often through the activation and repression of genes. Signals may induce transitory or permanent changes in cells.

Chemical signals specifically bind protein receptors found either on the plasma membrane or in the cell's cytoplasm. The signal pathways consist of a few nonprotein **second messengers** such as calcium ions (Ca^{2+}), cyclic adenosine monophosphate (cAMP), cyclic guanosine monophosphate (cGMP), diacylglycerol (DG) and inositol triphosphate

98

(IP$_3$) which **transduce** or send the signal to the cellular components involved in the response.

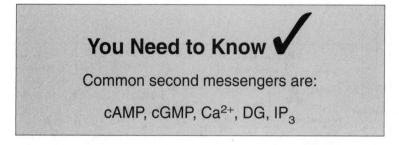

G Proteins

G proteins are important intermediates in signal transduction pathways because they determine whether the signal will be stimulatory or inhibitory. A major family of G proteins is **trimeric**, consisting of three subunits: α, β, and γ. The α subunit is capable of binding GDP or GTP. When a signal stimulates the receptor, the altered receptor stimulates a change in the G protein: GDP dissociates from the α subunit, and GTP takes its place. This stimulates dissociation of the α subunit, which then diffuses along the inner surface of the membrane until it contacts an enzyme or ion pore. The activity of the α subunit is blocked when the bound GTP is hydrolyzed and it reassociates with the β and γ subunits.

Trimeric G proteins affect ion pores and enzymes such as adenylcyclases, guanylcyclases, and phospholipases. The pores may be opened or closed, and the enzymes may be stimulated or inhibited. These enzymes are important in signal pathways because they amplify weak signals by catalytically producing second messengers.

A signal transduction pathway involving IP$_3$ and DG is shown in Figure 11-1. A growth factor or hormone binding to a cell-membrane receptor alters the receptor's conformation, which stimulates the dissociation of a neighboring trimeric G protein and a GDP attached to the α subunit of the G protein. The α subunit becomes active in the signal transduction pathway by dissociating from the β and γ subunits of the G protein and exchanging a molecule of GTP for GDP. The active G protein stimulates a membrane bound phospholipase C (PLC) that hydrolyzes the phosphatidylinositol 4',5'-bisphosphate (PIP$_2$) in the membrane to DG and inositol 1',4',5'-triphosphate (IP$_3$). IP$_3$ binding to calcium ion pores opens

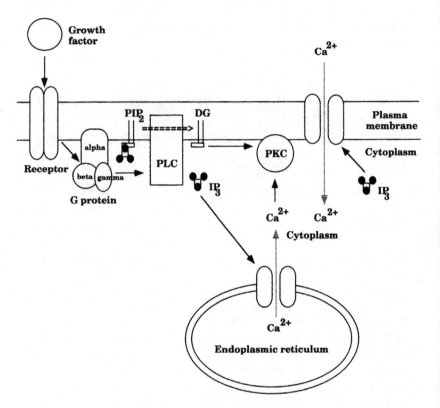

Figure 11-1 A signal transduction pathway.

these pores in the ER and the plasma membrane, allowing calcium ions to move along their concentration gradient from the ER and from the extracellular environment into the cytoplasm. Calcium ions and DG binding to inactive protein kinase C (PKC) causes PKC to become active. Activated PKC phosphorylates other protein kinases in signal transduction pathways, often activating them.

A second family of G proteins consist of a single subunit. These **monomeric** proteins are known as **Ras proteins** and are activated indirectly through autophosphorylation of membrane-bound tyrosine kinases and the regulatory proteins that interact with the phosphates (see Figure 11-2). The relative amounts of active and inactive Ras are determined by guanine nucleotide release factors (GNRFs) and by GTPase-activat-

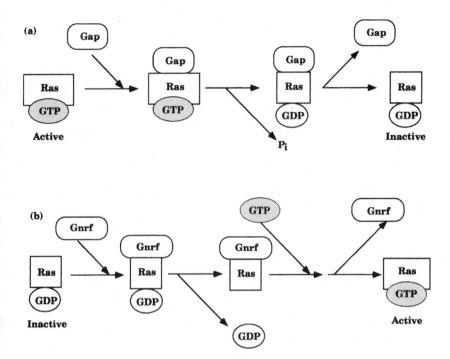

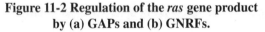

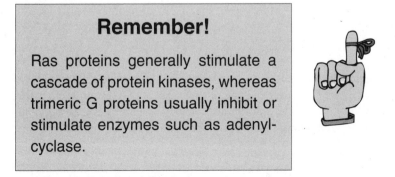

**Figure 11-2 Regulation of the *ras* gene product
by (a) GAPs and (b) GNRFs.**

ing proteins (GAPs). Since these proteins promote the exchange of GTP
for GDP, or GDP for GTP, respectfully, they affect Ras protein activity.
Hydrolysis of GTP to GDP and P$_i$ inhibits Ras. Some Ras proteins are
negatively regulated by tumor suppressor proteins.

Remember!

Ras proteins generally stimulate a
cascade of protein kinases, whereas
trimeric G proteins usually inhibit or
stimulate enzymes such as adenyl-
cyclase.

Kinases and Phosphatases

Protein kinases phosphorylate other proteins. Some are stimulated at the beginning of signal transduction pathways by specific growth factors, whereas others are stimulated at later points during the signal pathway by the binding of second messengers or by phosphorylation. Some kinases are membrane-bound, but the majority are free in the cytoplasm. Most protein kinases are multimeric, consisting of separate catalytic and regulatory subunits.

Protein kinases are categorized on the basis of which amino acids they phosphorylate (e.g., tyrosine and serine-threonine) as well as on the basis of their activity (e.g., cAMP-dependent, cyclin-dependent, etc.). Membrane tyrosine kinases and serine-threonine kinases are generally stimulated directly by a chemical signal. Protein kinase A (PKA) and protein kinase G (PKG) are soluble serine-threonine kinases activated by cAMP and cGMP, respectively. PKC is used to label a large family of serine-threonine protein kinases that are stimulated directly by the second messengers DG and/or Ca^{2+}. A protein kinase that requires the calcium ion binding protein **calmodulin** and Ca^{2+} for its activity is known as calmodulin-calcium-dependent protein kinase. Ca^{2+} pores in the plasma membrane and in the endoplasmic reticulum are opened by the binding of IP_3 to the pores. A protein kinase that is required for progression through the cell cycle is dependent upon a number of protein stimulators called **cyclins**.

Because protein kinases are part of most signal pathways, they control almost every aspect of cellular physiology. Thus, control of a cell's physiology is affected by the phosphorylated state of its proteins. If phosphorylation affects a response, then there must be enzymes that reverse that response. **Protein phosphatases** remove phosphate groups from proteins. Some phosphatases are activated by phosphorylation, some by a calmodulin-Ca^{2+} complex, and some by inhibitory proteins.

The Cell Cycle

Nonproliferating cells are in the "no growth" (G_0) stage of the cell cycle. G_0 cells are expressing only those genes that are needed to maintain life and carry out any specialized function they normally have. To begin the first growth stage (G_1) of the cell cycle, cells must be signaled, usually by more than one growth factor. The binding of growth factors to their receptors may stimulate the receptor's tyrosine kinase activities, resulting in autophosphorylation and in the phosphorylation of other kinases. Most of these secondary kinases are serine-threonine kinases. In some cases, activated receptors stimulate G proteins that subsequently stimulate enzymes that produce second messengers. The second messengers often stimulate serine-threonine kinases. The serine-threonine kinases phosphorylate transcriptional activators or repressors, stimulating and inhibiting them, respectively. This results in the expression of a number of genes, in particular, the genes for G_1 phase cyclins: CDPKs, RNA polymerase, DNA helicase, and DNA polymerase. The accumulation of these proteins during G_1 is essential for DNA replication in the S phase of mitosis.

G_1 phase cyclins are regulatory subunits for proteins called **mitosis-promoting factors** (MPFs). The catalytic subunits are designated Cdc (for cell division control), but are also known as Cdks (for cyclin-dependent kinases). The cyclin-Cdk complex is activated by phosphorylation and dephosphorylation of Cdk on certain threonines and tyrosines, which ultimately leads to DNA replication.

✸ Recognize these!

Cyclin-dependent protein kinase (CDPK) = MPF = CDC.

The second growth phase (G_2) occurs after DNA synthesis. G_2 cyclins begin to accumulate, increasing the activity of a new group of cell division control kinases that are activated by a sequence of phosphorylations and dephosphorylations. These kinases phosphorylate proteins called **lamins** in the nuclear membrane, scaffold proteins in the nucleus, and a microtubule associated protein kinase.

Phosphorylation of a chromosomal scaffold protein (believed to be a topoisomerase) results in the condensation of the solenoid chromosome structure onto a scaffold. Phosphorylation of lamins cause chromosomal detachment from the inner nuclear membrane and fragmentation of the nuclear membrane. Other phosphorylation events lead to tubulin polymerization necessary for mitotic divsion to ensue.

Solved Problems

Solved Problem 11.1 What are GAPs and GNRFs? What are their roles in cellular communication?

GAPs are GTP-activating proteins that inactivate Ras. GNRFs are guanine nucleotide release factors that stimulate Ras. Ras proteins are monomeric G protein intermediates in signal transduction pathways which involve numerous phosphorylation events.

Solved Problem 11.2 Chapter 10 discussed the relevance of oncoviruses in the development of cancer. How do signal transduction pathways become permanently altered and cells become transformed in the absence of viruses?

Spontaneous mutations (see Chapter 6) in various proto-oncogenes can convert them into oncogenes that may affect signal transduction pathways. Some mutations make signal transduction proteins nonfunctional, whereas others modify those domains inolved in protein regulation so that the signal transduction proteins are permanently in the active state. Cells that produce signal transduction proteins that are always active behave inappropriately. They may cause the cell to repeatedly progress through successive cycles. Mutations in the proteins that negatively control signal transduction may also lead to cellular transformation.

Once a cell has acquired a mutation in a proto-oncogene or tumor-suppressor gene that stimulates the cell's proliferation, secondary mutations in other proto-oncogenes or tumor suppressor genes occur that deregulate cells in the mutant population. These secondary mutations promote the survival and proliferation of many types of deregulated cells.

Chapter 12
MOLECULAR EVOLUTION

IN THIS CHAPTER:

✔ *Early Beginnings*
✔ *The RNA World*
✔ *The DNA World*
✔ *Phylogenetic Analysis*
✔ *The Evolution of Eukaryotic Cells*
✔ *Solved Problems*

Early Beginnings

Molecular studies have shed light on the origin of life and its subsequent evolution into a myriad of extinct and extant species. Theories regarding these early events are impossible to prove conclusively with circumstantial evidence. **Molecular fossils** such as introns within transcriptional units and common biochemcial pathways shared between diverse organisms provide additional support for current models.

Living cells possess: (1) a boundary membrane separating the cell's contents from its external environment; (2) one or more DNA molecules that carry genetic information for specifying the structure of proteins involved in replication of its own DNA, in metabolism, in growth, and in cell division; (3) a transcriptional system whereby RNAs are synthesized; (4) a translational system for decoding ribonucleotide sequences into amino acid sequences; and (5) a metabolic system that provides usable forms of energy to carry out these essential activities.

The first living system(s) were undoubtedly much simpler than any cells alive today. The transition from nonliving to living was gradual, and no single event led to life in all its modern complexity. Even today biologists cannot agree on a definition of life. The following criteria are usually included in attempts to define life. An aggregate of cells is considered "alive" if it (1) can use chemical energy or radiant energy to drive energy-requiring chemical reactions; (2) can increase its mass by controlled synthesis; and (3) possesses an information coding system and a system for translating the coded information into molecules that maintain the system and allow it to reproduce one or more collections of molecules with similar properties.

The best estimate for the age of the earth is 4.6 billion years. The oldest **microfossils** superficially resembling bacteria have been dated at about 3.5 billion years ago. Thus, **chemical evolution** (e.g., abiotic syntheses of amino acids and their polymerization into peptides) during the first 1.0 to 1.5 billion years of earth history probably preceded the appearance of cellular life and its subsequent **biological evolution**.

The major opinion is that earth's atmosphere was nearly neutral, nonoxidizing, and contained primarily nitrogen, carbon dioxide, hydrogen sulfide, and water. Microfossils resembling modern cyanobacteria ("blue-green algae") have been found in limestone rocks called **stromalites** dated 3.5 billion years ago. Presumably these ancient photosynthetic bacteria produced oxygen as a by-product of splitting water, just as cyanobacteria do today. Over more than another billion years, oxygen slowly began to accumulate, eventually causing the primitive atmosphere to become oxidizing.

Remember!

Early atmospheric conditions:

Hot
Neutral
Nonoxidizing
N, CO_2, H_2S, H_2O

There are two major scientific theories regarding how life came to be on earth. It either evolved on earth from nonliving chemicals, or it evolved elsewhere in the universe and was brought to the earth by comets or meteorites (**panspermia theory**). The belief that life was created by a supernatural force is impossible to support or refute with factual evidence and hence is outside the realm of science.

Amino acids and other precursors of modern biomacromolecules have been found inside meteorites, so chemical evolution of these molecules might have been (and still may be) widespread in the cosmos. In 1953, Stanley Miller, at the suggestion of his mentor Harold Urey, used a reflux apparatus to simulate early atmospheric conditions in an attempt to reproduce the chemical evolution of biological precursor molecules. He recirculated water vapor and other gases (CH_4, NH_3, and H_2) through a chamber where they were exposed to a continuous high voltage electrical discharge that simulated natural lightning. After a few days, the mixture was analyzed and found to contain at least ten different amino acids, some aldehydes, and hydrogen cyanide. Subsequent experiments by Miller and other researchers using different molecular mixtures and energy sources produced a variety of other building blocks of biological polymers.

Sidney Fox and his colleagues heated amino acids under water-free (anhydric) conditions to temperatures of 160–210°C and found amino acids polymerized into proteinlike chains which he called **proteinoids**, which are branched rather than linear. When dissolved in water, these proteinoids exhibit several properties of biological proteins including limited enzymatic activity and susceptibility to digestion by proteases.

Proteinlike peptides can also be synthesized from amino acids on clays. Clays consist of alternating layers of inorganic ions and water molecules. The highly ordered lattice structure of clays strongly attracts organic molecules and promotes chemical reactions between them. Polypeptides have been detected in laboratory simulations of these processes.

When solutions of proteinoids are heated in water and then allowed to cool, small, spherical particles called **microspheres** are formed. These microspheres are about the same size and shape as spherical bacteria. Some are able to grow (add mass) by accretion of proteinoids and lipids and subsequently proliferate by binary fission or budding.

Lipids can spontaneously organize into double-layered bubbles called **liposomes**, which are leaky enough to absorb various substances such as proteins from the surrounding medium. Substances trapped within the liposome find themselves in a hydrophobic environment that might provide more favorable conditions for certain kinds of chemical reactions. Thus, lipid bilayers may have promoted both aggregation and catalysis. Vesicles composed of lipid membranes and protein microspheres, but devoid of RNA or DNA molecules, are hypothesized to have existed in the early stages of life. These entities are called **progenotes**.

The RNA World

A living system must be able to replicate its genetic material and be capable of evolving. Proteins are necessary for DNA replication, but most proteins are synthesized on RNA templates that themselves were synthesized on DNA templates.

It has been hypothesized that RNA molecules capable of self-replication arose prebiotically by random condensation of mononucleotides into small polymers. The active sites of most modern proteins and catalytic RNAs constitute relatively small segments of the polymers to which they belong. The smaller primitive RNA replicase polymers, formed abiotically, would probably have only weak catalytic activity, and would have been subject to error-prone replication. But such a molecule might have been able to use itself or other RNA molecules as a template for polymerizing RNA nucleotides. The many errors made during replication of the early RNA replicase would create a pool of genetic diversity on which natural selection could act to favor those molecules that were able to replicate faster and/or have greater accuracy. One problem, however, is that no replicase can copy its own active site. It is thus necessary to propose that a minimum of two RNA replicases were synthesized at nearly the same time from the "primordial soup" of precursors. A primitive type of cell containing an RNA genome, called the **eugenote**, is hypothesized to have evolved from the progenote population.

RNA molecules were probably the primordial genome/enzyme molecules of primitive living systems. Ribose sugars are easier to synthesize under simulated primordial conditions than deoxyribose sugars. The DNA precursors of all extant cells are produced by reduction of RNA nucleoside diphosphates by the highly conserved protein enzyme **ribonucleoside diphosphate reductase**. This enzyme appears in all modern

cells with few structural differences, suggesting that
it is an ancient one that has performed the same essential task over a long evolutionary history. Living
systems with RNA genomes are presumed to have
evolved first. More stable DNA genomes evolved
later to store genetic information.

Also, ssDNA would have been less likely to
form complex three-dimensional configurations due to the lack of the 2′
hydroxyl, which may participate in unusual hydrogen bonds. Furthermore, the catalytic activity of some modern ribozymes is known to involve this 2′ OH. Lastly, dsDNA molecules have the same unvarying
double-helical structure that would not lead us to expect them to have enzymatic properties. However, they can fold back on themselves and
ssDNA can fold into tertiary structures.

 **Note!**

ssDNA molecules that cut RNA molecules can be
evolved through artificial selection in cell-free systems.

Gradually, proteins took over many of the catalytic functions origi-
nally performed by RNA molecules. This would have allowed for greater
flexibility in the sequences since there are 20 amino acids and only 4 ribonucleotides. Also, three-dimensional shapes in RNA molecules would
require a complementary sequence elsewhere on the strand to form hydrogen bonds.

Early life systems that could make a variety of useful proteins would
tend to have a selective advantage over those that had a more restric-
tive repertoire. Selection would thus promote the early protoribosomes,
tRNAs, and tRNA synthetases to diversify. This process is envisioned to
have produced a set of peptide-specific ribosomes, each with a different
internal guide sequence serving as an mRNA sequence. A primitive genetic code would thus become established as sets of tRNA synthetases
and peptide-specific protoribosomes evolved.

The DNA World

Double-stranded DNA molecules are more stable than ssRNA. It would thus be advantageous for living systems to store heritable information in DNA molecules rather than RNA molecules. The 2′ OH of RNA can attack an adjacent phosphodiester bond, rendering RNAs much more labile than DNAs. This autocatalytic process may have been accelerated by the harsh conditions on the primitive earth. As cells became more complex, their genome sizes had to increase. If early eugenotes had segmented RNA genomes, at least one of each segment would have to be present in each daughter cell for its survival. To enhance the probability that progeny cells are provided with a full genome, natural selection would favor production of polycistronic genomes. But the larger the RNA genomic segments are, the less stable they would become because of autocatalysis. Thus, it would be advantageous for more stable polycistronic DNA molecules to take over genomic functions of RNA, leaving the RNAs to carry out functions that need not require long-lived molecules. The earliest anucleate cells containing DNA genomes (and all subsequent such cells) are known as **prokaryotes**.

At least four major processes were required to complete this transition: (1) synthesis of DNA monomers by ribonucleoside diphosphate reductase; (2) reverse transcription of DNA polymers from RNA genomes; (3) replication of DNA genomes by a DNA polymerase; and (4) transcription of DNA genomes in functional (nongenomic) RNA molecules such as tRNA, mRNA, and rRNA.

The split genes of modern eukaryotic cells consist of coding regions (exons) and noncoding regions (introns). The interruption of the gene provided by introns offers an evolutionary advantage. Apparently, exons from different genes can sometimes be recombined by natural mechanisms to code for proteins of different functions but containing related amino acid domains. Each of these domains may have a specific function (e.g., binding to a receptor, forming an α-helix, etc.). This process, termed **exon shuffling**, is inferred to have been used extensively in the DNA world of early eukaryotes.

Phylogenetic Analysis

Proteins evolve at different rates because of intrinsic factors (repair mechanisms) and extrinsic factors (environmental mutagens). Highly conserved proteins apparently have only been able to tolerate a few minor changes, whereas some other proteins have been able to absorb many mutations without loss of function. Mutations that occur outside a region involved with normal function of the molecule may be tolerated as a selectively **neutral mutation**. Over geological time, these neutral mutations tend to accumulate within a geneological lineage. If it is assumed that such neutral mutations accumulate at a fairly constant rate for a highly conserved protein, it is possible to establish the branching pattern of a **phylogenetic tree** (also called a **cladogram** or an **evolutionary tree**).

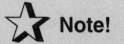 **Note!**

Some evolution rates (point mutations per 100 million years):

Triose phosphate isomerase = 3
Hemoglobin = 21
Nonfunctional pseudogenes = 400

The **principle of parsimony** is commonly used to determine the minimum number of genetic changes required to account for the amino or nucleotide sequence differences between organisms sharing a common ancestor. The evolutionary distances separating organisms in a phylogenetic tree are usually expressed in units of nucleotide mutations or amino acid substitutions along each arm of the tree between branch points (see Figure 12-1).

The Evolution of Eukaryotic Cells

At one time, prokaryotes were thought to be more closely related to a postulated progenote (the common ancestor of all cells, before there was a

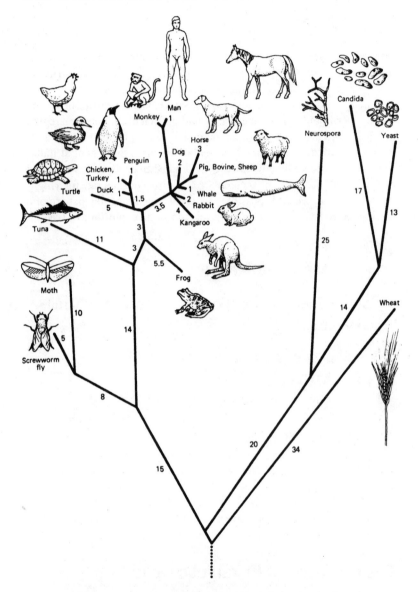

Figure 12-1 A phylogenetic tree based on homologies between cytochrome *c* molecules in various organisms. Branch length is represented by the most likely number of point mutations that occurred during evolution of these species.

genome) than were eukaryotes, and all prokaryotes were thought to be more closely related to one another than to any eukaryote. Most prokaryotic species can be further classified as **eubacteria**. The other prokaryotic subkingdom, **archae**, occupies the kinds of environments that were presumed to be widespread when life first evolved. Hence, it was commonly believed that eubacteria evolved from primitive archae and eukaryotes evolved from eubacteria. Gradually, however, many more differences were found to separate the two subkingdoms. Some archae traits are shared with the eubacteria (they are both prokaryotes), whereas others are shared with eukaryotes (e.g., genes for rRNAs and tRNAs contain introns). Based on his analysis of nucleotide sequences in the highly conserved 16S rRNAs from many organisms, Carl Woese proposed in 1977 that archae are as different from eubacteria as either group is from the eukaryotes. Today, it is thought that all three lines have descended from the same progenote.

Organisms with a nucleus may have evolved as long ago as 3.5 billion years, but how the first nuclear membrane arose remains a mystery. According to the **membrane proliferation hypothesis**, one or more invaginations of the plasma membrane in the progenote coalesced internally to surround the genome, became severed from the plasma membrane, and formed a double-layered nuclear membrane. The manner of infolding of the plasma membrane shown in Figure 12-2 accounts for the fact that the nucleus of modern eukaryotic cells is enclosed within a "double membrane" consisting of two lipid bilayers. Note that a portion of the ER is continuous with the outer membrane of the nuclear envelope.

The origin of mitochondria in younger eukaryotes may be explained by the **endosymbiotic theory**. Some ancient cells were capable of ingesting food particles by endocytic invaginations of their plasma membranes. It is possible that at least one large, fermenting, feeder cell engulfed one or more smaller respiratory bacteria, but failed to digest them. This **endosymbiont** was able to survive in an environment where nutrients were abundant and it could hide from other predatory cells. In turn, the host feeder cell gained the energetic advantages of oxidative respiration over fermentation (see Chapter 1). These complementary advantages evolved into a **symbiotic** ("living together") relationship wherein neither entity can survive without the other. Part of this mutual adaptation in-

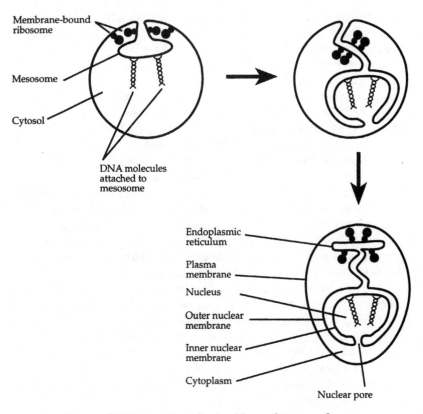

Figure 12-2 Formation of a double nuclear membrane.

volved the transfer of most of the genes of the bacterial endosymbiont into the nucleus of the host cell. Most negatively charged molecules, including mRNAs, tRNAs, rRNAs, and some proteins, that cannot cross the membrane of these organelles must still be encoded by the genomes of these organelles. This process is proposed to have given rise to the mitochondria of modern eukaryotic cells at least 1.5 billion years ago.

⭐ **Note!**

A type of purple, photosynthetic bacteria that had lost its photosynthetic ability and retained its respiratory chain is hypothesized to represent the endocytosed bacteria.

A stronger case can be made for the evolution of chloroplasts by endosymbiosis than that for mitochondria. An aerobic, eukaryotic feeder cell (one that had already evolved mitochondria) is proposed to have engulfed one or more eubacteria (related to cyanobacteria) that were capable of oxygenic photosynthesis. In the process of evolving into chloroplasts, the endosymbionts relinquished some of their genes to the nuclear genome, but not as many as did the endosymbionts that evolved into mitochondria. Like the mitochondria, the protochloroplasts had to retain all of the genes specifying tRNAs and rRNAs for protein synthesis within the chloroplast.

Much evidence supports the endosymbiotic theory for the origin of chloroplasts and mitochondria. These organelles are approximately the same size as bacteria. The genomes reside within a single, circular DNA molecule that is devoid of histone proteins, like bacteria. Both organelles reproduce asexually by growth and division of existing organelles in a manner similar to binary fission. Protein synthesis in mitochondria and chloroplasts is inhibited by a variety of antibiotics that inactivate bacterial ribosomes, but have little effect on cytoplasmic eukaryotic ribosomes. Nascent polypeptides in bacteria, mitochondria, and chloroplasts have N-formylmethionine at their amino ends. Mitochondria and chloroplast genomes encode the tRNA and rRNA molecules for their own protein-synthesizing systems. The ribosomes in both organelles resemble bacterial ribosomes in size and structure. Lastly, the endosymbiotic theory accounts for the fact that both organelles have double membranes. The inner membrane corresponds to the plasma membrane of the ancestral endosymbiont; the outer membrane represents the plasma membrane of the ancestral feeder host cell.

Interesting

One theory suggests that the flagella and cilia of eukaryotes originated from motile, symbiotic bacteria on the surface of ancestral eukaryotic cells.

Solved Problems

Solved Problem 12.1 Would you imagine introns are relatively new features of genomes or that they were present in early forms of life?

Today introns are abundant in genomes of vertebrates, less frequent in lower eukaryotes, and absent from all common bacteria. But if introns were present in primitive genomes, most bacteria and relatively simple eukaryotes might have lost them under selection pressure to streamline their genomes for more rapid reproduction at lower energy expenditure. If, on the other hand, introns were not present in early genomes, but were inserted by recombination mechanisms into more advanced genomes, the simpler organisms may have resisted this process. However, random insertion events would most likely destroy the encoding of essential amino acids, rather than preserve them. Analysis of ancient, ubiquitous, highly conserved proteins may help resolve this problem.

Solved Problem 12.2 What are the advantages of using nucleotide sequences in constructing phylogenies rather than amino acid sequences?

Nucleotide sequencing is much faster and less expensive than peptide sequencing. Even tiny amounts of DNA in fossils over 100 million years old have been successfully sequenced by using PCR to amplify the DNA. There is no comparable method for multiplying tiny bits of polypeptides to levels needed for sequencing.

In addition, DNA sequences can reveal silent mutations, whereas protein analyses cannot. Furthermore, DNA analyses are not restricted to sequences coding for proteins, but can also be used for genes that encode tRNAs and rRNAs, as well as noncoding control sequences, introns, spacers, or any part of the genome.

Solved Problem 12.3 What function does a nuclear membrane serve that would give that cell a selective advantage over an anucleate cell?

The nuclear membrane keeps ribosomes and many other large cytoplasmic molecules confined to the cytosol. Primary mRNA transcripts from split genes must undergo several kinds of processing including the removal of introns and splicing of exons before they are released from the nucleus to the cytoplasm for translation into proteins. Without a nuclear membrane to separate ribosomes from pre-mRNA, many translated proteins would contain amino acid sequences of introns that had not yet been spliced out. This might also result in shortened proteins if ribosomes encountered a stop codon within an intron.

Index